Lecture Notes in Control and Information Sciences

Volume 463

About this Series

This series aims to report new developments in the fields of control and information sciences—quickly, informally and at a high level. The type of material considered for publication includes:

1. Preliminary drafts of monographs and advanced textbooks
2. Lectures on a new field, or presenting a new angle on a classical field
3. Research reports
4. Reports of meetings, provided they are
   (a) of exceptional interest and
   (b) devoted to a specific topic. The timeliness of subject material is very important.

More information about this series at http://www.springer.com/series/642

Masayoshi Toda

Robust Motion Control of Oscillatory-Base Manipulators

$\mathcal{H}_\infty$-Control and Sliding-Mode-Control-Based Approaches

Springer

Masayoshi Toda
Department of Ocean Sciences
Tokyo University of Marine Science
and Technology
Tokyo
Japan

Additional material to this book can be downloaded from http://extras.springer.com.

ISSN 0170-8643 ISSN 1610-7411 (electronic)
Lecture Notes in Control and Information Sciences
ISBN 978-3-319-21779-6 ISBN 978-3-319-21780-2 (eBook)
DOI 10.1007/978-3-319-21780-2

Library of Congress Control Number: 2015944735

Mathematics Subject Classification: 70Q05

Springer Cham Heidelberg New York Dordrecht London

Printed on acid-free paper

Springer International Publishing AG Switzerland is part of Springer Science+Business Media (www.springer.com)

Preface

The subject of the monograph is robust control of "An oscillatory-base manipulator (OBM)" which can be regarded as a model system for a variety of mechanical systems installed on the oscillatory base. Further, this class of systems can be divided into subclasses, one of which mainly includes offshore mechanical systems, e.g., offshore cranes, installed on the oscillatory base being affected by wave-induced disturbances, and a typical example of the other subclass is a space robot system mounted on the flexible base which tends to oscillate due to the intrinsic flexible structure. This monograph will focus on the former subclass, i.e., offshore mechanical systems.

The monograph introduces such control problems, and presents some control methodologies to solve them which the author has developed and demonstrates, with respect to control system design and analysis, particularly on "robustness" and control performances by simulations and hardware experiments. The common feature of such control problems can be stated as "how to achieve successful tracking control in the presence of disturbances due to the base oscillation and further model uncertainties and variations." Therefore, the model of an oscillatory-base manipulator is a very important control objective because it serves as a tool to solve various control application problems in the marine industries, but also can contribute to the control science community as a benchmark system of disturbance rejection and robust tracking control.

The control methodologies presented in the monograph are based on $\mathcal{H}_\infty$ control and sliding-mode control, both of which are now well-known powerful control schemes to solve robust control problems. The author believes that the control problems considered are interesting and useful examples with respect to applications of those well-known control schemes. Therefore, the author strongly hope that not only control engineers involved in marine systems, but also engineers in the other areas and students in the course of control science get interested in such problems, and control theory scientists will employ the problems to evaluate their own novel control schemes.

Please note that the monograph has been designed to be as self-contained as possible. Additionally, the reader can download some demonstration programs for MATLAB® from http://www2.kaiyodai.ac.jp/~toda/obm/.

The author would like to express his sincere gratitude to Springer and the staff who have given the author such an opportunity to publish the monograph and helped a lot. And, the author wishes to express his thanks to his students who were involved in this project and made the achievements together with the author, particularly, Mr. Masahiro Sato who supported him by preparing the estimation algorithm-based programs in Chap. 7. Finally, the author wishes to express his deep gratitude to his wife and children for their encouragement, support, and their patience during the preparation of the monograph.

Tokyo, Japan
June 2015

Masayoshi Toda

Contents

Notation

Abbreviations

DOF	Degree of freedom
FFT	Fast Fourier transform
LFT	Linear fractional transformation
LQ	Linear quadratic
MIMO	Multi-input multi-output
OBM	Oscillatory-base manipulator
PID	Proportional integral derivative
PSD	Power spectral density
RMSE	Root-mean-square error
RSSI	Rotating sliding surface with variable-gain integral control
SDC	State-dependent coefficient
SISO	Single-input single-output
SMC	Sliding-mode control or controller
TDOF	Two-degree-of-freedom
VSS	Variable structure system

Symbols

$\|\| \cdot \|\|$	Euclidean norm
$\|\| \cdot \|\|_F$	Frobenius norm
$\|\| \cdot \|\|_\infty$	$\mathcal{H}_\infty norm$
$(\cdot)^T$	Transpose
$\mathbb{R}$	The field of real numbers
$\mathbb{R}^n$	The n-dimensional real vector space
$\mathbb{R}^{n \times m}$	The n by m real matrix space
$\mathbf{C}$	The field of complex numbers
$\mathbf{C}^{n \times m}$	The n by m complex matrix space
$\mathcal{H}_\infty$	The stable transfer function matrix space

$\succ 0$	Positive definite
$\prec 0$	Negative definite
I	The identity matrix
I_n	The n by n identity matrix
$\mathrm{diag}(\cdot)$	Diagonal matrix
$\mathrm{blockdiag}(\cdot)$	Block diagonal matrix
$\bar{\sigma}(\cdot)$	The largest singular value
$\mathcal{F}_l(\cdot,\cdot)$	Lower LFT
$\mathcal{F}_u(\cdot,\cdot)$	Upper LFT
$\mathcal{S}(\cdot,\cdot)$	Redheffer star-product
S	The sensitivity function matrix
T	The complementary sensitivity function matrix
T_a	The quasi-complementary sensitivity function matrix
μ	The structured singular value
μ_c	The μ with constant scalings
$(\cdot)_{\mathrm{n}}$	Nominal parameter
$\mathcal{P}(\cdot)$	Orthogonal projection
$vec(\cdot)$	Mapping from $\mathbb{R}^{2\times 2}$ to $\mathbb{R}^4$
$vec^{-1}(\cdot)$	Inverse mapping of *vec*
ϱ_1	The criterion 1 for the hyperplane
ϱ_2	The criterion 2 for the hyperplane
$\bar{e}$	Root-mean-square error (RMSE)

Chapter 1
Introduction

Abstract This introductory chapter first defines the key terminology "oscillatory-base manipulator," which represents a model system associated with mechanical systems installed on the oscillatory base. This category consists of two groups depending on the presence or the absence of external oscillatory disturbance. Typical examples with external oscillatory disturbance can be found in marine mechanical systems, such as offshore cranes, drill ships. The other group, i.e., with no external oscillatory disturbance, mainly contains space robots mounted on the flexible base. This monograph will focus on the former group of marine mechanical systems. We introduce related works and our own research history where the $\mathcal{H}_\infty$ control framework and the sliding-mode control one have played important roles, and present the organization of the monograph.

1.1 What Is an Oscillatory-Base Manipulator?

"An oscillatory-base manipulator (OBM)" which we will focus on throughout this monograph is a model system of a variety of mechanical systems installed on the oscillatory base. In particular, typical examples can be found in the marine engineering and science fields, such as offshore cranes, drill ships, marine observatory systems, on-board radar gimbal systems, and so on, which are subject to wave-induced base oscillation. In this monograph, we will address motion control problems for oscillatory-base manipulators by analyzing their dynamics, developing control design methods, and demonstrating our proposed control systems.

The common feature of such control problems can be stated as "how to achieve successful tracking control in the presence of disturbances due to the base oscillation and further model uncertainties and variations." Additionally, the desired trajectory for the system to track is strongly affected by the base oscillation in many control application scenarios. Therefore, the model of an oscillatory-base manipulator is a very important control objective because it serves not only as a tool to solve various control application problems in the marine industries but also as a benchmark system of disturbance rejection and robust tracking control for the control science

M. Toda, *Robust Motion Control of Oscillatory-Base Manipulators*,
Lecture Notes in Control and Information Sciences 463,
DOI 10.1007/978-3-319-21780-2_1

community. Due to the nature of the problems, additional requirements are necessary for the control systems for OBMs compared to the conventional robot control systems, e.g., [1, 8, 19, 53, 77, 89–91].

In general, control problems of OBMs can be categorized into two classes depending on the presence or absence of external oscillatory disturbance. One class of them contains problems in the presence of external oscillatory disturbance which is the main cause of base oscillation, to which offshore mechanical systems as described above and mobile robots with manipulators running on a rough road belong [35, 36, 46, 64, 110]. The other class is of control problems in the absence of such external disturbance, where, due to the intrinsic flexible structure of the system, motion of the manipulator induces the base oscillation, for instance, a large space robot mounted on a flexible base or free-flying one maintaining a space satellite is an example. The latter class of problems have gained a lot of research interests in the space engineering field. Manipulator systems involved in these works are referred to as "macro/micro manipulators" [18, 74, 87, 99]. The central issues of such problems are first to decouple the manipulator and base dynamics and second to decay the base oscillation. To satisfy those requirements, several approaches have been proposed such as task space feedback [36, 87], filtering command [65], path-planning [70, 79], acceleration feedback [57], active damping [58], $\mathcal{H}_\infty$ control [96, 97], Fuzzy control [61], pseudo-inverse jacobian [62]-based approaches.

It should be noted that these two types of OBM control problems are essentially different ones due to difference in the mechanisms of base oscillation arising. Therefore, requirements for the respective control systems are also different. In the former problems, the control system is expected to cope with the external oscillatory disturbance, in other words, to be able to excite the system according to the disturbance. In theory, so called the "internal model principle" comes in, and the requirement for the linear time-invariant control system is equivalent to having the corresponding poles to the frequencies of disturbance. On the other hand, in the latter problems, conversely the control system need not to excite the intrinsic oscillatory modes. If the control system is linear time-invariant one, this requirement can be achieved by letting the controller have the corresponding zeros to the oscillatory modes. In [97], we have realized the requirement by utilizing $\mathcal{H}_\infty$ control. Interestingly enough, these two problems form a contrast to each other, in terms of system poles and zeros for the controllers.

In this monograph, we focus on the former class of problems and furthermore consider robust control against model uncertainties due to the payload variations, which is often the case for oceanic industries such as the fishing industry and the ocean mining industry where a variety of objects in size, shape, and mass are to be handled. In recent years, there have been increasing requirements for offshore mechanical systems such as high operability, accuracy, efficiency, and safety of operation, as the role of the ocean has becoming more and more important in terms of energy, mineral, and food resources, and further the global environmental problems.However, offshore environments are considerably difficult for mechanical systems because of wave, tide, and wind, which make the control problems of offshore mechanical systems challenging.

Here, we introduce some of the previous works on control of offshore mechanical systems related to the problem of OBMs. This class of control problems of OBMs contains two types of problems according to the coordinate systems, that is, base-fixed coordinates (local coordinates) and earth-fixed coordinates (global coordinates). Most of the previous works considered global coordinates. As the most popular control problem in such problems considering global coordinates, heave (vertical) motion compensation systems have been extensively studied aiming at applications in ocean platforms for deep water oil exploitation [2, 7], drill-ships [15, 52], offshore cranes [21, 38, 45, 54, 71, 73], and remotely operated vehicles [34]. The operations of the control systems in the ocean engineering literature can be classified into two types, one of which is an operation to reduce oscillations of ships or ocean platforms [2, 7], and the other one is to compensate their oscillatory motions [15, 21, 38, 45, 54, 52, 67, 71, 73]. The control operation considered in the monograph belongs to the latter type.

Then, we address the related works from the viewpoint of control theory. Being based on the assumption that the dynamical model is almost linear, a PD-control-based approach has been presented in [52], and LQ optimal control has been employed in [2, 7, 38]. In [45], targeting on wire velocity control of an offshore crane, a method of generating a reference signal for wave synchronization has been proposed using a transfer function model. Skaare and Egeland [88] have presented a parallel force/position control scheme with a similar transfer function model to that in [45]. Do and Pan [15] have developed a nonlinear control system with a disturbance observer and a back-stepping technique. Neupert et al. [71] have exploited the scheme of disturbance decoupling in [40] for a nonlinear model of a offshore crane. In [67], a nonlinear internal-model-based approach has been developed for autonomous vertical landing of an aircraft, whose dynamics is not affected by oscillatory disturbances. Except [52], in these papers, their proposed control systems have been evaluated by performing numerical simulations as in [2, 7, 15, 38, 67] and/or real model experiments [45, 71, 88].

The common objectives in such control problems of compensating oscillatory motions are first tracking control according to the desired trajectory, which is strongly influenced by the base oscillation, and second suppressing disturbance due to the base oscillation. It is important to note that these control objectives can be interpreted into the frequency-domain control objectives as long as the system can be regarded as a linear time-invariant one, and the frequency range of the base oscillation can be known in advance. In fact, which is the case for an offshore OBM, and hence we considered application of $\mathcal{H}_\infty$ control to such problems. $\mathcal{H}_\infty$ is well known to be useful and effective for such frequency-domain control objectives and moreover robust control problems.

1.2 Previous Works and Contents of the Monograph

First, we began with the simplest problem of an OBM in [94], where an $\mathcal{H}_\infty$-control-based approach to motion control in local coordinates, i.e., base-fixed coordinates, was examined for a one degree-of-freedom (DOF) manipulator with a one-DOF base, thus a single-input single-output (SISO) system. The key idea in [94] as the control methodology was collaboration of $\mathcal{H}_\infty$ control, linear state-feedback control, and nonlinear state-feedback control. This method was extended to problems of multiple-DOF manipulators in local coordinates, i.e., a multi-input multi-output (MIMO) system [95] and further its robust control feature was enhanced by developing a machinery, the "extended matrix polytope" and by utilizing μ-synthesis based on the machinery. Then, in [82], we applied the method to the global-coordinate problems of OBMs by extending the control design method for local-coordinate problems. Until the work [82], in order to evaluate the proposed method, we had performed simulations and confirmed its effectiveness. After that, we developed an hardware experimental apparatus which consists of a two-DOF manipulator and a one-DOF base. In [83], we improved the method by employing a new weighting function for $\mathcal{H}_\infty$ control design, which not only reduces influence of sensor error but also enhances the robustness of the control systems; furthermore the proposed method was evaluated by simulations and hardware experiments. The results showed that the proposed control system can successfully perform in tracking control while suppressing disturbance due to the base oscillation, further even in the presence of model uncertainties due to the payload variations.

Meanwhile, we attempted to apply sliding-mode control (SMC), which is also known to be powerful to disturbance rejection and robust control problems [5, 20, 39, 41, 60, 73, 102, 103, 111]. In [42], we proposed a novel nonlinear sliding surface with the variable-gain integral control. The proposed SMC provides superior performance to the conventional SMC in terms of transient performance and less required control inputs. The advantage of the SMC-based approach over the $\mathcal{H}_\infty$-control-based one is that it does not require information of frequencies of the base oscillation. On the other hand, its disadvantages are common shortcomings of SMC, that is, the gaps between in theory and in practical applications mainly due to discontinuous switching control inputs. In [42], only preliminary results have been given.

In those works, we assumed that accurate measurements of the base oscillation are available for the control system. In [84], we considered estimation problem of the base oscillation via a low-cost gyro rate sensor, and proposed a method of selectively combining multiple $\mathcal{H}_\infty$ filters. The important idea in [84] is a new criterion to select appropriate filters, which is based on innovations. The simulation results showed that the algorithm is useful but still needs to be improved.

By being based on our research history of robust control of OBMs and adding new research results, we have completed this monograph. In this monograph, we have reorganized and integrated the proposed methodologies and given a new perspective to them. The chapters of the monograph are organized as follows.

- Chap. 2 presents the problem definition, including important assumptions, the dynamical model formulation, and the base oscillation models, on which we will be based throughout the monograph.
- Chap. 3 addresses the experimental apparatus which consists of a two-DOF manipulator and a one-DOF base. With this apparatus, we are able to conduct robust control tests by changing the attached payloads of 11 different weights. We also analyze how these payload variations result in parametric variations in the dynamical model of the manipulator.
- Chap. 4 presents the key content in our research history, that is, the $\mathcal{H}_\infty$-control-based approach. First, we introduce our machinery the "extended matrix polytope" to efficiently and less conservatively represent model uncertainties due to the payload variations. Then, we explain and demonstrate the control design method. Finally, we analyze the designed controllers with respect to the fundamental properties of the controller, i.e., poles and zeros, and frequency responses, and further robustness.
- Chap. 5 demonstrates evaluations on the designed controllers in Chap. 4 by simulations and experiments, which include robust control tests and comparison with the conventional PID controller.
- Chap. 6 presents the SMC-based approach and evaluates the method by simulations, which also include robust control tests and comparison with the proposed $\mathcal{H}_\infty$-control-based approach.
- Chap. 7 introduces the estimation algorithm for the base oscillation, which deploys a novel idea of selectively combining multiple $\mathcal{H}_\infty$ filters. By simulations, we demonstrate its estimation performance.

Chapter 2
Problem Definition, Dynamical Model Formulation

Abstract This chapter presents the problem definition and dynamical model formulation of an oscillatory-base manipulator considering an illustrative example, which we will work on throughout the monograph. We will consider three types of control problems, specifically attitude control in local coordinates (base-fixed coordinates) which is associated with on-board operations such as cargo handling, attitude control in global coordinates (earth-fixed coordinates), e.g., radar gimbal systems, and position control in global coordinates, e.g., heave-motion-compensated cranes. Further, as patterns of base oscillation, we consider three patterns, single-frequency sinusoidal oscillation, double-frequency sinusoidal one, and ocean-wave imitated oscillation based on the Bretschneider spectrum. Using combinations of those cases, we will demonstrate control system design and analysis, control simulations and experiments.

2.1 Introduction

First, this chapter defines motion control problems of an OBM which we will work on throughout this monograph, by choosing an illustrative model and setting some important assumptions. Then, the problems are categorized into two classes, local-coordinate problems and global-coordinate ones, depending on the coordinate frame to be referred by the control system. We consider three types of control problems. Subsequently, being based on the problem definition, the corresponding dynamical model is derived. Further, we introduce three patterns of base oscillation which we will use for various demonstrations together with the control problems.

2.2 Problem Definition

As an illustrative model of OBMs, we consider a two-DOF manipulator with a one-DOF oscillatory base and a payload as depicted in Fig. 2.1, and motions of which are restricted on the vertical plane. The reason that this simple model has been

M. Toda, *Robust Motion Control of Oscillatory-Base Manipulators*,
Lecture Notes in Control and Information Sciences 463,
DOI 10.1007/978-3-319-21780-2_2

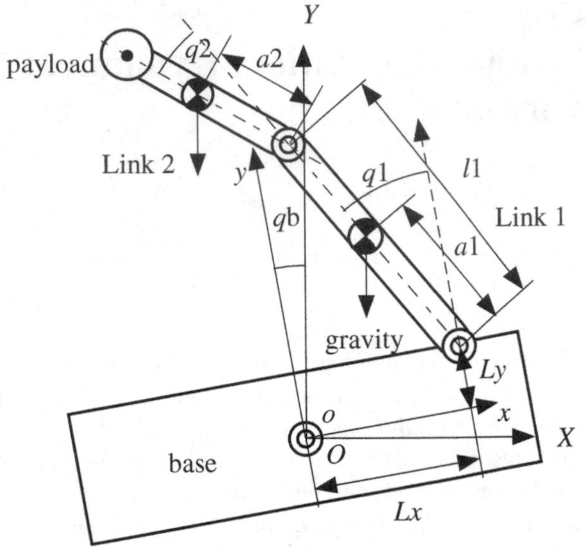

Fig. 2.1 Schematic diagram of the illustrative model of OBMs (a two-DOF manipulator with a one-DOF oscillatory base and a payload)

employed is because, in practical cases of offshore mechanical systems, roll and heave (vertical) oscillations among six-DOF ones are manifest and critical. It should be noted that the roll motion of the model with respect to the origin is one-DOF rotating one, however, with respect to another point, e.g., the root of the manipulator, involves heave and sway motions as well as roll ones. Further, from the viewpoint of control, forces parallel and torques orthogonal to the rotating axes of manipulator joints do not affect the manipulator dynamics. Therefore, this illustrative model is simple for exposition, but is realistic enough. Further, notice that a payload is attached to the tip of the second link in Fig. 2.1, which is assumed to have uncertainty in its physical parameters, e.g., mass and inertia moment, in order to analyze robust control problems.

Here the following important assumptions are made in addressing motion control problems of OBMs;

- A1 the frequency range of the base motion is known in advance;
- A2 the forces and torques exerted on the base by the manipulator are negligible to the base motion;
- A3 the actuator dynamics is negligible;
- A4 the payload has uncertainty in its physical parameters, e.g., mass and inertia moment;
- A5 except the payload, the physical parameters of the manipulator are known;
- A6 the joint angles and velocities of the manipulator can be measured; and
- A7 the base oscillation angle can be measured.

A1 is natural and is not restrictive considering the ocean environment and, further, is very important to make our proposed $\mathcal{H}_\infty$-control-based technique highly effective.

In fact it has been reported that the frequencies of ocean surface waves are within the range 1/30–1 Hz [9]. A2, which supports A1 together with the facts in oceanography and also makes the control problem simple, shall be discussed later in detail and be clarified not to be restrictive, but to be reasonable. A3 is only for simplicity. A4 is intended to analyze robust control problems, and is often the case for offshore mechanical systems such as a crane handling mineral and fishery resources. A5 and A6 are necessary for control systems we will demonstrate and standard conditions in terms of motion control. These assumptions are taken into account in Chaps. 2–6, and Chap. 7 which presents an estimation method of the base oscillation aims at implementing A7.

Taking into account all the above conditions, now the motion control problem of OBMs to be dealt with can be stated as "*under the assumptions A1–A7, synthesize an controller that achieves successful motion control of the manipulator in the presence of disturbance due to the base oscillation and model uncertainties in the payload physical parameters.*"

2.3 Local-Coordinate and Global-Coordinate Problems

See that in Fig. 2.1 two different coordinate frames are set; OXY is an inertia frame whose Y–axis is parallel to the direction of gravity and oxy is a frame attached to the base whose origin o is fixed at O. Motion control problems of OBMs can be typically categorized into two classes depending on the coordinate frame to be referred by the control system. One class is of the case where the control system performs being based on a base-fixed coordinate frame, i.e., a local-coordinate frame (as oxy in Fig. 2.1), which is called the class of *local-coordinate problems*. The other class contains *global-coordinate problems* where an earth-fixed coordinate frame, i.e., an inertia frame (as OXY in Fig. 2.1), is referred. The local-coordinate problems are associated with practical applications with base-fixed task spaces, such as a ship-mounted crane performing load/unload operations of cargo on the ship. On the other hand, an on-board radar gimbals and a ship-mounted crane performing on land fall into the class of global-coordinate problems.

The common feature between both the problems in global and local coordinates is the disturbances due to the base oscillation. On the contrary, the difference between them is in that in global coordinates the desired trajectory of motion must be generated according to the base motion, while it is not the case in local coordinates where the reference signals do not contain information on the base oscillation and further there is no need to measure the base motion as shown later.

Moreover, regardless of coordinate frames and in general, motion control problems are divided into *position control* ones and *attitude control* ones. In the case of OBMs, "position control" implies that the position of the payload is to be controlled, while "attitude control" does controlling the attitudes of the links. Hence, by combining the coordinate frame classes with these motion control types, four patterns of motion control problems of OBMs can be considered as depicted Figs. 2.2, 2.3, 2.4

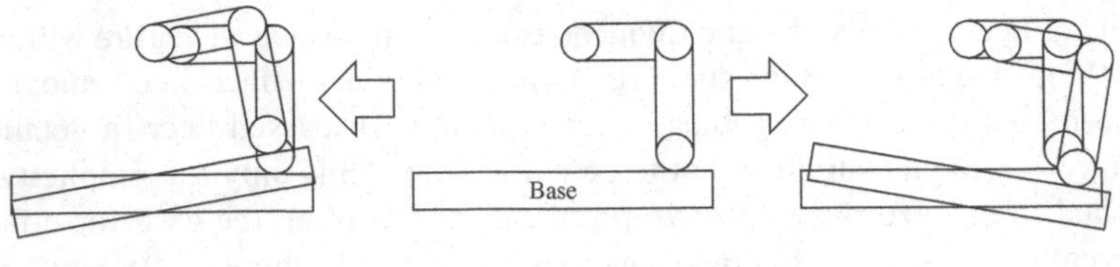

Fig. 2.2 Position control in a base-fixed coordinate system (local case)

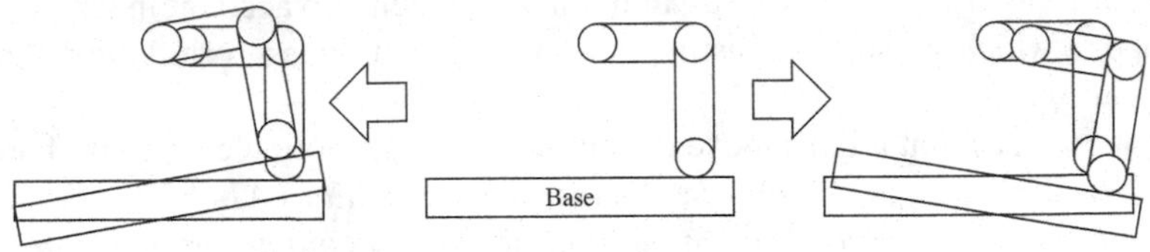

Fig. 2.3 Attitude control in a base-fixed coordinate system (local case)

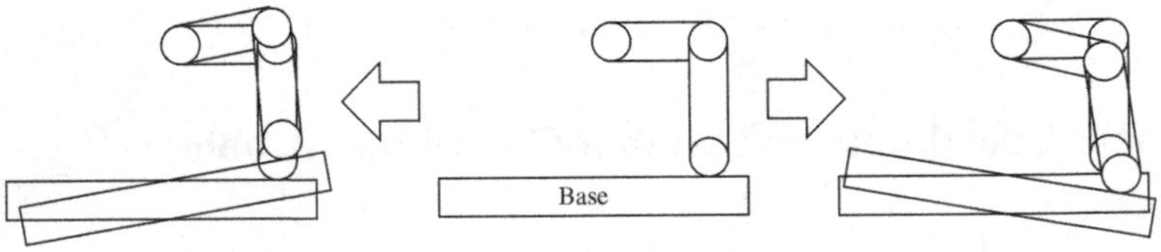

Fig. 2.4 Position control in a global-coordinate system

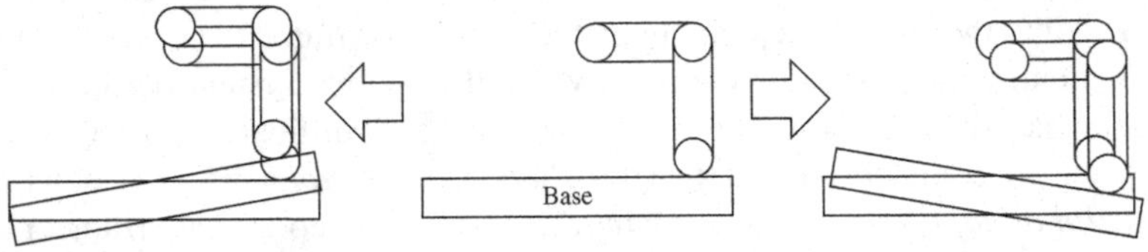

Fig. 2.5 Attitude control in a global-coordinate system

and 2.5. As seen from the figures, notice that, in local-coordinate problems, position control, and attitude control are equivalent as long as the redundancy and/or multiplicity of solution of the joint space mapped to the given task space can be ignored, whereas it is not the case for global-coordinate problems.

2.4 Dynamical Model Formulation

2.4.1 Dynamical Model of an OBM

The dynamics of n-rigid link manipulators with revolutionary joints subject to m-DOF base oscillation can be described by the standard formulation for manipulator dynamics plus the disturbance due to the base oscillation as follows:

$$M(\boldsymbol{q})\ddot{\boldsymbol{q}} + C(\boldsymbol{q},\dot{\boldsymbol{q}})\dot{\boldsymbol{q}} + D\dot{\boldsymbol{q}} + G(\boldsymbol{q},\boldsymbol{q}_\mathrm{b}) + H(\boldsymbol{q},\dot{\boldsymbol{q}},\dot{\boldsymbol{q}}_\mathrm{b},\ddot{\boldsymbol{q}}_\mathrm{b}) = \boldsymbol{\tau}, \tag{2.1}$$

where $\boldsymbol{q} \in \mathbb{R}^n$ and $\boldsymbol{q}_\mathrm{b} \in \mathbb{R}^m$ denote the position vectors of the manipulator links and the base, respectively, and $\boldsymbol{\tau}$ is the input torque vector; $M(\boldsymbol{q}) \in \mathbb{R}^{n\times n}$ is the inertia matrix of the manipulator, which is a symmetric positive definite matrix; $C(\boldsymbol{q},\dot{\boldsymbol{q}})\dot{\boldsymbol{q}} \in \mathbb{R}^n$ represents the centripetal and Coriolis torque depending only on the states of the manipulator; $D \in \mathbb{R}^{n\times n}$ is the damping coefficient matrix of the manipulator joints, which is a positive definite constant diagonal matrix; $G(\boldsymbol{q},\boldsymbol{q}_\mathrm{b}) \in \mathbb{R}^n$ is the gravitational torque; $H(\boldsymbol{q},\dot{\boldsymbol{q}},\dot{\boldsymbol{q}}_\mathrm{b},\ddot{\boldsymbol{q}}_\mathrm{b}) \in \mathbb{R}^n$ represents the inertia torque and the centripetal and Coriolis torque due to the base oscillation. $G(\boldsymbol{q},\boldsymbol{q}_\mathrm{b})$ and $H(\boldsymbol{q},\dot{\boldsymbol{q}},\dot{\boldsymbol{q}}_\mathrm{b},\ddot{\boldsymbol{q}}_\mathrm{b})$ form the disturbance prescribed, which are nonlinearly coupled with both the states of the manipulator and the base.

Now, we derive the specific formulation of (2.1) for the illustrative model in Fig. 2.1 by applying Lagrangian mechanics. The notations used here are as follows.

q_b, q_1, q_2	Position angles of the base and the links as defined in Fig. 2.1
J_1, J_2	Inertia moment of each link with respect to the centroid
m_1, m_2	Mass of each link
D_1, D_2	Damping coefficient of each joint
L_x, L_y, l_1, a_1, a_2	Geometric parameters as defined in Fig. 2.1
τ_1, τ_2	Control torque applied to each joint
g	Gravitational acceleration
t	Time variable

Then, each term in the model (2.1) can be explicitly represented by (2.2)–(2.6) and the base oscillation is described by (2.7); $(\cdot)^T$ denotes the transpose.

$$\begin{aligned}
M(\boldsymbol{q}) &= \begin{bmatrix} M_{11} & M_{12} \\ M_{21} & M_{22} \end{bmatrix} \\
M_{11} &= m_1a_1^2 + m_2(a_2^2 + l_1^2 + 2a_2l_1\cos(q_2)) + J_1 + J_2 \\
M_{12} &= M_{21} = m_2(a_2^2 + a_2l_1\cos(q_2)) + J_2 \\
M_{22} &= m_2a_2^2 + J_2,
\end{aligned} \tag{2.2}$$

$$C(\boldsymbol{q},\dot{\boldsymbol{q}})\dot{\boldsymbol{q}} = \begin{bmatrix} -m_2a_2l_1\sin(q_2)\dot{q}_2(2\dot{q}_1+\dot{q}_2) \\ m_2a_2l_1\sin(q_2)\dot{q}_1^2 \end{bmatrix}, \tag{2.3}$$

$$D = \mathrm{diag}[D_1, D_2], \tag{2.4}$$

$$\begin{aligned}
G(\boldsymbol{q},\boldsymbol{q}_\mathrm{b}) &= [G_1, G_2]^T \\
G_1 &= -\{m_1a_1\sin(q_b+q_1) + m_2(l_1\sin(q_b+q_1) \\
&\quad +a_2\sin(q_\mathrm{b}+q_1+q_2))\}g \\
G_2 &= -m_2a_2\sin(q_\mathrm{b}+q_1+q_2)g,
\end{aligned} \tag{2.5}$$

$$
\begin{aligned}
H(\boldsymbol{q}, \dot{\boldsymbol{q}}, \dot{\boldsymbol{q}}_\mathrm{b}, \ddot{\boldsymbol{q}}_\mathrm{b}) &= [H_1, H_2]^T \\
H_1 &= \{m_1(a_1{}^2 - a_1 L_x \sin(q_1) + a_1 L_y \cos(q_1)) \\
&\quad + m_2(a_2^2 + l_1{}^2 - l_1 L_x \sin(q_1) + l_1 L_y \cos(q_1) \\
&\quad - a_2 L_x \sin(q_1 + q_2) + a_2 L_y \cos(q_1 + q_2) \\
&\quad + 2 l_1 a_2 \cos(q_2)) + J_1 + J_2\}\ddot{q}_\mathrm{b} \\
&\quad + \{m_1(a_1 L_x \cos(q_1) + a_1 L_y \sin(q_1)) \\
&\quad + m_2(l_1 L_x \cos q_1 + l_1 L_y \sin q_1 \\
&\quad + a_2 L_x \cos(q_1 + q_2) + a_2 L_y \sin(q_1 + q_2))\}\dot{q}_\mathrm{b}^2 \\
&\quad - 2 m_2 a_2 l_1 \sin(q_2)\dot{q}_\mathrm{b}\dot{q}_2 \\
H_2 &= \{m_2(a_2^2 + a_2 l_1 \cos(q_2) - a_2 L_x \sin(q_1 + q_2) \\
&\quad + a_2 L_y \cos(q_1 + q_2)) + J_2\}\ddot{q}_\mathrm{b} \\
&\quad + m_2(a_2 l_1 \sin(q_2) + a_2 L_x \cos(q_1 + q_2) \\
&\quad + a_2 L_y \sin(q_1 + q_2))\dot{q}_\mathrm{b}^2 \\
&\quad + 2 m_2 a_2 l_1 \sin(q_2)\dot{q}_\mathrm{b}\dot{q}_1.
\end{aligned} \tag{2.6}
$$

$$
q_\mathrm{b} = \Sigma_{i=1}^{n_\omega} A_{\omega i} \sin(\omega_i t + \phi_i) \tag{2.7}
$$

As in (2.7), the base oscillation is modeled as a linear combination of multiple sinusoidal motions. Equation (2.6) indicates that the manipulator is strongly influenced by the base oscillation and that the larger amplitude and angular frequency of the base will induce the larger $H(\boldsymbol{q}, \dot{\boldsymbol{q}}, \dot{\boldsymbol{q}}_\mathrm{b}, \ddot{\boldsymbol{q}}_\mathrm{b})$. Hence, to achieve desirable motion control, how to overcome $H(\boldsymbol{q}, \dot{\boldsymbol{q}}, \dot{\boldsymbol{q}}_\mathrm{b}, \ddot{\boldsymbol{q}}_\mathrm{b})$ is the central issue.

2.4.2 Base Oscillation Model

In this monograph, we have chosen three types of base oscillations in (2.7) for demonstrations of control design and control system evaluation. The first one is a single-frequency one with $\omega_1 = 2\pi$ (rad/s), $A_{\omega 1} = 10^{(\circ)}$, and $\phi_1 = 0$ (rad), and the second one is a double-frequency one with $\omega_1 = \pi$, $\omega_2 = 2\pi$ (rad/s), $A_{\omega 1} = A_{\omega 2} = 5^{(\circ)}$, and $\phi_1 = \phi_2 = 0$ (rad). Furthermore, for more realistic demonstrations, continuously distributed frequencies imitating an ocean-wave spectrum are also considered for the third oscillation model. As in [15], we have employed *the two-parameter Bretschneider spectrum* [22], and by approximating the spectrum each component of base motion in (2.7) is generated as in the following.

$$
\omega_i = \omega_{min} + \frac{\omega_{max} - \omega_{min}}{n_\omega - 1}(i-1) \quad (i = 1, 2, \cdots, n_\omega) \tag{2.8}
$$

$$
S_{\omega i} = \frac{1.25}{4}\frac{\omega_0^4}{\omega_i^5} A_s^2 e^{-1.25(\omega_0/\omega_i)^4} \tag{2.9}
$$

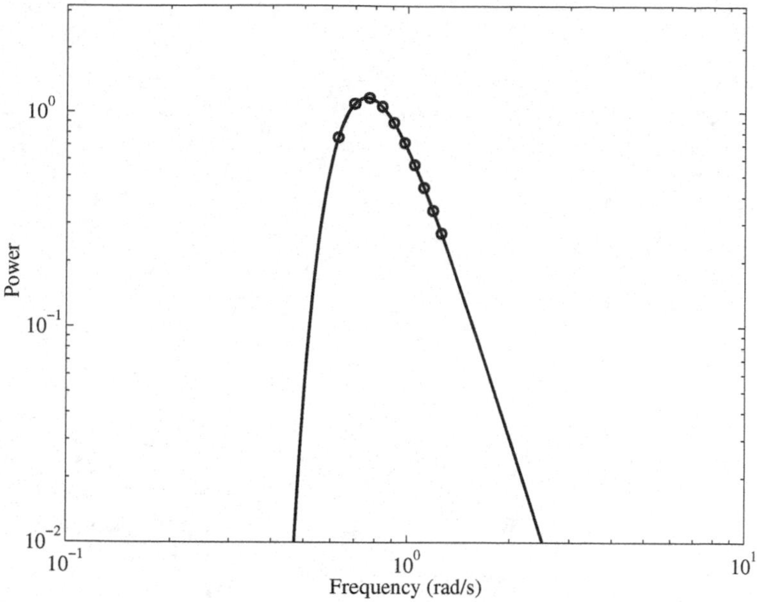

Fig. 2.6 Power spectrum of the Bretschneider model

$$A_{\omega i} = \frac{\omega_i^2}{9.8}\sqrt{2S_{\omega i}\frac{\omega_{max} - \omega_{min}}{n_\omega - 1}} \tag{2.10}$$

where ω_0, ω_{min}, and ω_{max} are the modal, minimum, and maximum frequencies, respectively, and A_s is the significant roll amplitude. Given these parameters, the spectral density $S_{\omega i}$ and amplitude A_i can be obtained. The phase ϕ_i is given as a random number between 0 and 2π with the uniform distribution.

To generate a base rolling motion, those parameters are set as $\omega_0 = 0.24\pi$, $\omega_{min} = 0.2\pi$, $\omega_{max} = 0.4\pi$ (rad/s), $n_\omega = 10$, $A_s = \pi$ (rad). With those parameters, the power spectrum is displayed in Fig. 2.6, where the circles represent the employed frequencies. Moreover, since our experimental apparatus is a small-scale model, taking into account the scale effect, (2.7) is modified as

$$q_b(t) = \Sigma_{i=1}^{n_\omega} A_{\omega i} \sin(5\omega_i t + \phi_i), \tag{2.11}$$

i.e., the frequencies are magnified by 5 times so as to make the disturbance affect enough. We refer to the oscillation as *the Bretschneider oscillation* hereafter.

Chapter 3
Experimental Apparatus and Analysis on Parameter Variation Due to Payload

Abstract We developed an experimental apparatus in order to evaluate control systems for oscillatory-base manipulators (OBMs) by hardware experiments. We call the apparatus the "experimental OBM," which was designed to be in accordance with the problem definition described in Chap. 2, and moreover to accommodate robust control demonstrations by exchanging the attached payloads of different weights. In this chapter, we first introduce the experimental OBM. Then, considering the illustrative model given in Chap. 2, and the experimental OBM, we analyze how the parameters in the dynamical model vary according to the payload variations, the results of which will be utilized in robustness analyses and robust control simulations for the control systems in the sequel.

3.1 Introduction

This chapter introduces an apparatus that we developed for experimental evaluation on control systems for OBMs, which is called "experimental OBM." The experimental OBM was designed to accommodate payload variation assuming robust control problems. Utilizing the illustrative model in Chap. 2, and the experimental OBM, we analyze the parameter variation in the dynamical model of manipulator due to payload variation.

3.2 Experimental OBM

In order to perform experimental evaluations on control systems for OBMs, we have designed and developed an apparatus (experimental OBM) such that the illustrative model in Fig. 2.1 can be implemented. Figure 3.1 shows a photo of the experimental OBM which has the same configuration as that of the illustrative model in Fig. 2.1, namely a two-DOF manipulator with a one-DOF oscillatory base. By conducting some system identification experiments, we have obtained the physical parameters of the manipulator as shown in Table 3.1, *which are deployed for control designs,*

M. Toda, *Robust Motion Control of Oscillatory-Base Manipulators*,
Lecture Notes in Control and Information Sciences 463,
DOI 10.1007/978-3-319-21780-2_3

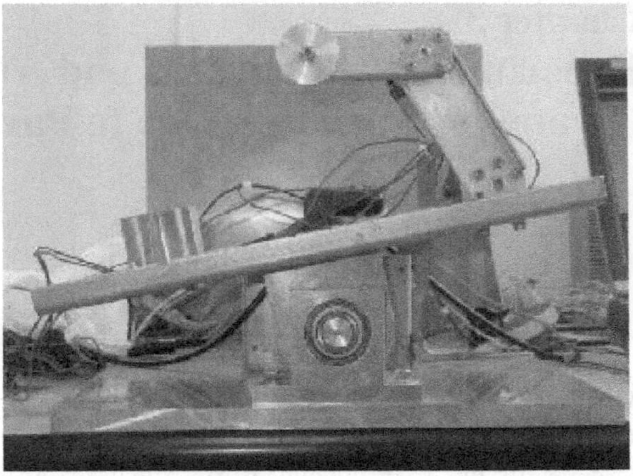

Fig. 3.1 Experimental oscillatory-base manipulator

Table 3.1 Physical parameters of the experimental OBM

Parameter	Value	Unit	Parameter	Value	Unit
J_1	7.373×10^{-4}	kg m^2	J_2	1.228×10^{-4} *	k gm^2
m_1	2.180×10^{-1}	kg	m_2	6.030×10^{-2} *	kg
a_1	5.739×10^{-2}	m	a_2	1.947×10^{-2}*	m
l_1	8.000×10^{-2}	m	l_2	6.000×10^{-2}	m
D_1	3.183×10^{-4}	N m s	D_2	6.074×10^{-5}	N m s
L_x	1.000×10^{-1}	m	L_y	6.000×10^{-2}	m
g	9.800	m/s^2			

* These values are ones of the nominal case with the payload width of 5 mm

Table 3.2 Specifications of the motors installed on the experimental OBM

	Motor b	Motor 1	Motor 2
Power (W)	250	90	4.5
Maximum continuous torque (mNm)	924	113	17
Maximum torque (mNm)	1165	262.5	26.9

control system analyses, and simulation demonstrations of control system performance throughout the monograph. The payload attached to the tip of Link 2 is exchangeable with other payloads of various widths, i.e., various masses, which is intended for analyses of robust control. In Table 3.1, the parameter values of Link 2 are ones of the nominal case with the payload width of 5 mm.

The joints are driven by the respective DC servo motors, Motor b for the base, Motor 1 for Link 1, and Motor 2 for Link 2, whose specifications are shown in Table 3.2. According to the specifications, the control torques are restricted as τ_1 within ± 0.2625 Nm and τ_2 within ± 0.0269 Nm. The angles of the links and base q_1,

q_2, and q_b are measured via the attached encoders with the resolution of 0.18 °, and the angular velocities are estimated using the backward difference of the angle data. The sampling period for measurement and control is 0.01 s.

3.3 Analysis on Parameter Variation Due to Payloads

This section analyzes parameter variation in the dynamical model of manipulator by utilizing the illustrative model shown in Fig. 2.1 and the experimental OBM. Note that this analysis is not exclusive for OBMs but for general manipulators with payload variations. As mentioned in the previous section, the payload attached to the experimental OBM is exchangeable. We have five types of payloads that have the same circular cross section and the different widths of 1, 2, ⋯, 5 mm, respectively. By choosing a single or multiple payloads, it is possible to implement various cases of payload mass and inertia moment. Figure 3.2 shows a photo of samples of the payloads and Table 3.3 presents their respective widths and masses.

We demonstrate how the parameters in the dynamical model in (2.2)–(2.5) vary due to changes of the payloads. The case of the payload with 5 mm width is regarded as the nominal case. To facilitate the analysis, we rewrite the dynamical model as follows.

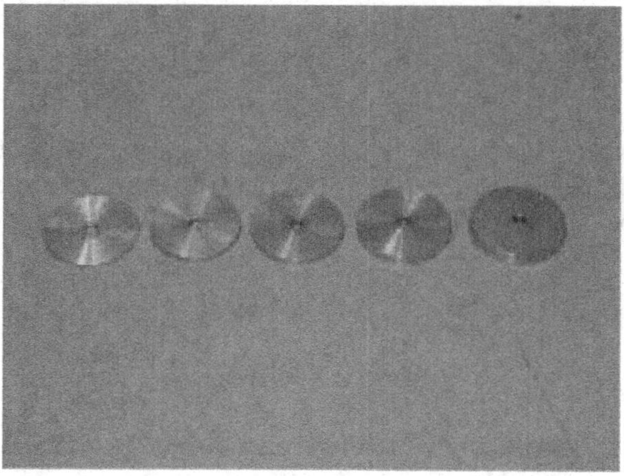

Fig. 3.2 Sample of payloads for the experimental OBM

Table 3.3 Widths and masses of the payloads for the experimental OBM

Width (mm)	1	2	3	4	5
Mass (g)	2.06	4.11	6.21	8.28	10.49

In the inertia matrix $M(\boldsymbol{q})$,

$$\begin{aligned}
M_{11} &= M_1 + 2R\cos(q_2),\\
M_{12} &= M_{21} = M_2 + R\cos(q_2),\\
M_{22} &= M_2,\\
M_1 &= m_1 a_1^2 + m_2(a_2^2 + l_1^2) + J_1 + J_2,\\
M_2 &= m_2 a_2^2 + J_2,\\
R &= m_2 a_2 l_1.
\end{aligned} \tag{3.1}$$

Hence, the centripetal and Coriolis term is rewritten as:

$$C(\boldsymbol{q}, \dot{\boldsymbol{q}})\dot{\boldsymbol{q}} = \begin{bmatrix} -R\sin(q_2)\dot{q}_2(2\dot{q}_1 + \dot{q}_2) \\ R\sin(q_2)\dot{q}_1^2 \end{bmatrix}. \tag{3.2}$$

The variation of mass of the payload from that in the nominal case is equal to the variation of mass of Link 2. Thus, let Δm_2 denote the variation of mass. With the subscript $(\cdot)_n$ which denotes the nominal parameter, $\Delta m_2 = m_2 - m_{2n}$. Then, considering the shape and configuration of the payload, the variations in a_2(position of centroid) and M_2 in (3.1) can be calculated by the following:

$$\Delta a_2 = \frac{\Delta m_2(l_2 - a_{2n})}{m_{2n} + \Delta m_2}, \tag{3.3}$$

$$\Delta M_2 = \Delta m_2\left(\frac{r_1^2 + r_2^2}{2} + l_2^2\right), \tag{3.4}$$

where r_1 is the inner radius and r_2 is the outer radius of the circular cross section of the payload. Further using (3.3) and (3.4), the variations in the other parameters can be obtained as follows:

$$\Delta M_1 = \Delta m_2\left(\frac{r_1^2 + r_2^2}{2} + l_1^2 + l_2^2\right), \tag{3.5}$$

$$\Delta R = \Delta m_2 l_1 l_2. \tag{3.6}$$

From the above, therefore, the variations in the respective terms in (2.2)–(2.5) are represented by

$$\begin{aligned}
\Delta M(\boldsymbol{q}) &= M(\boldsymbol{q}) - M_n(\boldsymbol{q})\\
&= \Delta m_2 M'(\boldsymbol{q})\\
&= \Delta m_2 \begin{bmatrix} M'_{11} & M'_{12} \\ M'_{21} & M'_{22} \end{bmatrix},\\
M'_{11} &= \frac{r_1^2 + r_2^2}{2} + l_1^2 + l_2^2 + 2l_1 l_2\cos(q_2),
\end{aligned}$$

Table 3.4 Parameter variations due to change of the payload

Payload width (mm)	Variation rate* (%)	m_2 ($\times 10^{-2}$kg)	a_2 ($\times 10^{-2}$m)	M_1 ($\times 10^{-4}$kg m^2)	M_2 ($\times 10^{-4}$kg m^2)	R ($\times 10^{-5}$kg m^2)
0	−17	4.981	1.093	1.881	1.065	4.36
1	−14	5.187	1.288	1.901	1.142	5.34
2	−11	5.392	1.467	1.922	1.219	6.33
3	−7	5.602	1.637	1.944	1.297	7.34
4	−4	5.809	1.793	1.964	1.374	8.33
5	0	6.030	1.947	1.987	1.457	9.39
6	3	6.236	2.081	2.008	1.533	10.38
7	7	6.441	2.206	2.029	1.610	11.37
8	10	6.651	2.325	2.050	1.688	12.37
9	14	6.858	2.436	2.071	1.765	13.36
10	17	7.079	2.548	2.093	1.848	14.43

* Variation rate of m_2

$$
\begin{aligned}
M'_{12} = M'_{21} &= \frac{r_1^2 + r_2^2}{2} + l_2^2 + l_1 l_2 \cos(q_2), \\
M'_{22} &= \frac{r_1^2 + r_2^2}{2} + l_2^2, \qquad (3.7)
\end{aligned}
$$

$$
\begin{aligned}
\Delta C(\boldsymbol{q}, \dot{\boldsymbol{q}})\dot{\boldsymbol{q}} &= C(\boldsymbol{q}, \dot{\boldsymbol{q}})\dot{\boldsymbol{q}} - C_n(\boldsymbol{q}, \dot{\boldsymbol{q}})\dot{\boldsymbol{q}} \\
&= \Delta m_2 \begin{bmatrix} -l_1 l_2 \sin(q_2)\dot{q}_2(2\dot{q}_1 + \dot{q}_2) \\ l_1 l_2 \sin(q_2)\dot{q}_1^2 \end{bmatrix}, \qquad (3.8)
\end{aligned}
$$

$$
\begin{aligned}
\Delta G(\boldsymbol{q}, \boldsymbol{q}_b) &= G(\boldsymbol{q}, \boldsymbol{q}_b) - G_n(\boldsymbol{q}, \boldsymbol{q}_b) \\
&= [\Delta G_1, \Delta G_2]^T, \\
\Delta G_1 &= -\Delta m_2\{l_1 \sin(q_b + q_1) + l_2 \sin(q_b + q_1 + q_2)\}g, \\
\Delta G_2 &= -\Delta m_2 l_2 \sin(q_b + q_1 + q_2)g, \qquad (3.9)
\end{aligned}
$$

where the terms with the subscript $(\cdot)_n$ denote the corresponding terms with the nominal parameters.

It should be noticed that all the variations of the respective terms in the manipulator dynamical model except the damping term is completely linear with respect to the variation of mass of the payload Δm_2. This remarkable feature stems from the special property and configuration of the payload considered here, that is,

- the inertia moment is proportional to the payload mass; and
- the position of centroid of the payload on the link is invariant.

However, even in practical cases similar situations might be expected. Hence, this feature can be and should be exploited in robust control design as demonstrated in the subsequent chapter. Table 3.4 shows the corresponding parameters to the respective payload widths, which are employed for demonstrations of control design, control system analysis, simulations, and experiments.

Chapter 4
Motion Control Using an $\mathcal{H}_\infty$-Control-Based Approach

Abstract This chapter presents the key contents of the monograph, that is, the $\mathcal{H}_\infty$-control-based approach for the OBM robust control problems. This control methodology consists of several schemes, specifically nonlinear state-feedback control to reduce the nonlinearity, parametric model uncertainty representation, $\mathcal{H}_\infty$ control with weighting functions, and additional linear state-feedback control to compensate for the $\mathcal{H}_\infty$ control scheme. In particular, in order to less conservatively and effectively represent parametric model uncertainties due to the payload variations, we have developed a machinery the "extended matrix polytope," which is an extension of the conventional matrix polytope, in the aim of representation of a non-convex parameter space. In this chapter, we begin with brief review of the basic notion of $\mathcal{H}_\infty$ control, and introduce the extended matrix polytope. Then, we present the control design method with some design examples considering four types of $\mathcal{H}_\infty$ controllers, and perform analyses on the designed control systems in terms of properties such as system poles and zeros, frequency response, and robustness by utilizing two different approaches, one of which is based on μ-analysis with the extended matrix polytope, and the other one is a Lyapunov-theory-based one with a state-dependent coefficient (SDC) form. From the results of analyses, the designed control system reveals favorable properties in terms of disturbance rejection, tracking control, and robustness. Hence, their practical control performances can be also expected, which will be shown in the next chapter.

4.1 Introduction

This chapter presents the $\mathcal{H}_\infty$-control-based approach for the OBM robust control problems aforementioned, which is the heart of the monograph. The characteristic features of the problem of OBMs are first disturbance due to the base oscillation; second model uncertainties due to the payload variation; and third the intrinsic nonlinearity of the robotic manipulator. The first feature can be interpreted into the frequency-dependent property. The second one requires a control design method with which one can cope with robustness. When these two features are accounted for, the $\mathcal{H}_\infty$ control framework can be naturally a powerful candidate. The $\mathcal{H}_\infty$

M. Toda, *Robust Motion Control of Oscillatory-Base Manipulators*,
Lecture Notes in Control and Information Sciences 463,
DOI 10.1007/978-3-319-21780-2_4

control that we have employed is fundamentally for linear time-invariant systems. However, by adding some modifications, the nonlinearity can be reduced and be absorbed into the model uncertainty.

Therefore, the point is how to accommodate and cope with the disturbance, model uncertainty, and nonlinearity. In this context, we have developed the method to express parametric uncertainties efficiently and effectively such as uncertainties due to the payload variation. This method was developed by improving "matrix polytope," thus we call the method "extended matrix polytope" which is a less conservative approach than the conventional one and naturally contains the conventional one. We will demonstrate the effectiveness of the extended matrix polytope in control design and robustness analysis.

This chapter is organized as follows. In the next section, we briefly review the fundamental notions and tools of $\mathcal{H}_\infty$ control and μ-analysis and synthesis, which will be minimum requisite for us to explain our idea in the sequel. Then, Sect. 4.3 introduce the extended matrix polytope. Further, Sect. 4.4 presents the control design method and analyzes the obtained control systems. Taking into account of the page length, evaluations on the designed control systems by simulations and experiments will be presented in the next chapter.

4.2 $\mathcal{H}_\infty$ Control and μ-Analysis and Synthesis

In this section, we briefly review the $\mathcal{H}_\infty$ control framework. The history of this attractive control design methodology started with the works [113, 114] by Zames in the early 1980s, which pointed out that, as criteria to evaluate control systems, the $\mathcal{H}_\infty$ norm of the transfer function matrix of a closed-loop control system is important. Before that, for instance, the well-developed linear quadratic (LQ) optimal control framework had been based on the $\mathcal{H}_2$ norm which represents a sort of average performance over the frequency domain. While, the $\mathcal{H}_\infty$ norm represents a sort of the maximal gain over the frequencies, and thus is strongly related to the worst-case performance. Then, this idea was connected with so-called "the small-gain theorem" and has developed to be one of the most effective and popular robust control design framework [50, 104]. By motivated Zames's works, a lot of research efforts have been devoted and provided important and useful achievements [16, 24, 28, 37, 44, 51, 80]. In principle, the $\mathcal{H}_\infty$ control method is available for linear, time-invariant systems. However, this methodology has been extendedly applied to nonlinear, time-varying systems by elaborate modifications, exactly as in this monograph, and, on the other hand, also in the theoretical field, the notion of $\mathcal{H}_\infty$ norm is extended to nonlinear system in a natural manner as the induced $\mathcal{L}_2$-gain-based control.

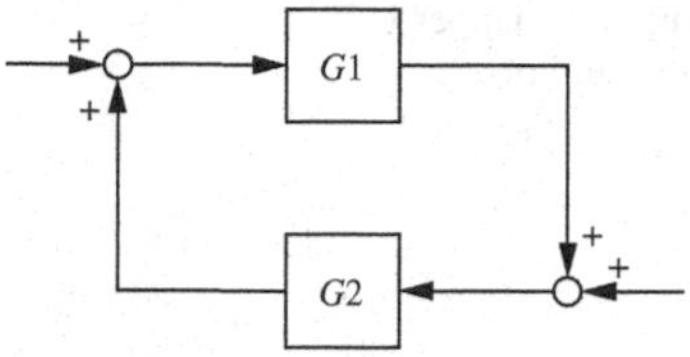

Fig. 4.1 Feedback configuration for the small-gain theorem

4.2.1 Small-Gain Theorem and Linear Fractional Transformations

Here, we address the considerably important theorem which made the $\mathcal{H}_\infty$ control framework a powerful robust control design method, that is, "the small-gain theorem". We consider a feedback configuration of two systems as depicted in Fig. 4.1. Let $G_1(s)$ and $G_2(s)$ denote the transfer function matrices of the respective systems in Fig. 4.1. Then, for the closed-loop system, the following theorem is known as the small-gain theorem.

Theorem 4.1 (Small-Gain Theorem) *[14, 30, 115] Suppose $G_1(s) \in \mathcal{H}_\infty$ and $G_2(s) \in \mathcal{H}_\infty$ in the feedback configuration in Fig. 4.1. If and only if*

$$||G_1(s)G_2(s)||_\infty < 1 \text{ and } ||\mathrm{G}_2(\mathrm{s})\mathrm{G}_1(\mathrm{s})||_\infty < 1, \tag{4.1}$$

the closed-loop system is internally stable.

In the robust control design approach, one of the two systems in Fig. 4.1 is designated as the uncertainty, and the other one plays a role of the interconnected system consisting of the plant, controller, even fictitious weighting functions for design. Then, in order to express such a configuration in design and analysis, the following formula, called "linear fractional transformation (LFT)", is useful and is commonly used.

Consider a block diagram in Fig. 4.2, which usually represents the plant including the uncertainty denoted by Δ in modeling and/or control design. The block M contains the nominal plant (i.e., when $\Delta = 0$) and maybe additional weighting functions reflecting the design specifications. Thus, M is in general referred to as a "generalized plant". M's input port for $\boldsymbol{u}$ and output port for $\boldsymbol{y}$ are usually used for a feedback controller to be designed. Let Δ and M represent the system matrices, then according to the respective dimensions M can be partitioned as

$$M = \begin{bmatrix} M_{11} & M_{12} \\ M_{21} & M_{22} \end{bmatrix}, \tag{4.2}$$

Further, If $(I - M_{11}\Delta)$ is invertible, the upper LFT [17, 30, 115] can be defined as

$$\mathcal{F}_u(M, \Delta) := M_{22} + M_{21}\Delta(I - M_{11}\Delta)^{-1}M_{12}, \tag{4.3}$$

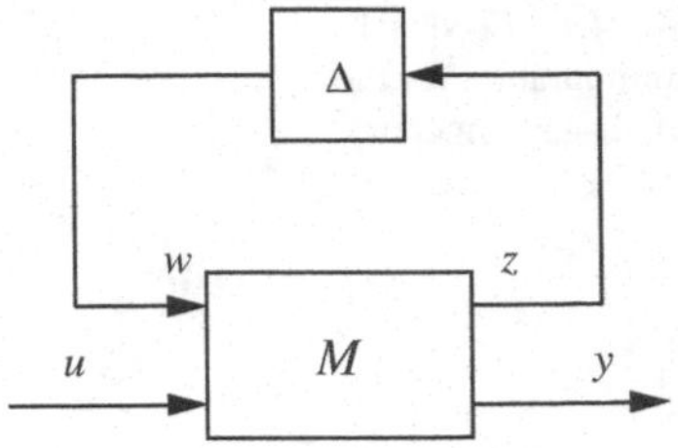

Fig. 4.2 Upper LFT configuration

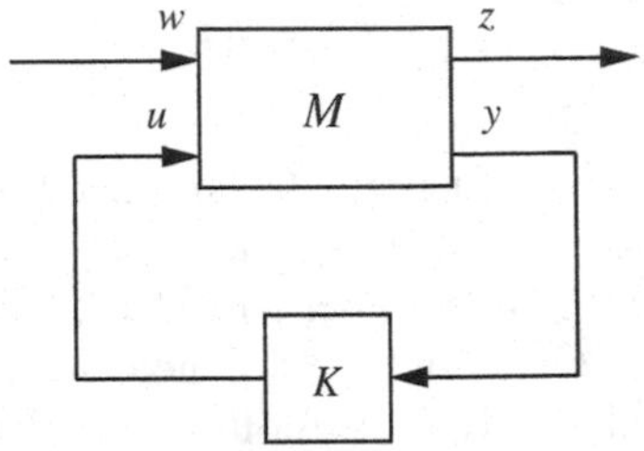

Fig. 4.3 Lower LFT configuration

which represents the transfer function matrix from u to y in Fig. 4.2.

On the other hand, one may consider a upside-down configuration as in Fig. 4.3 in robustness analysis and/or problem formulation for $\mathcal{H}_\infty$ control, where K represents the feedback controller. In this configuration, M's input port for $\boldsymbol{w}$ and output port for $\boldsymbol{z}$ are used to evaluate the robustness of the closed-loop control system. Being based on the same partitions of M in (4.2), if $(I - M_{22}K)$ is invertible, this configuration is given by

$$\mathcal{F}_l(M, K) := M_{11} + M_{12}K(I - M_{22}K)^{-1}M_{21}, \tag{4.4}$$

which represents the transfer function matrix from $\boldsymbol{w}$ to $\boldsymbol{z}$ in Fig. 4.3, and is called the lower LFT.

4.2.2 $\mathcal{H}_\infty$ Control Standard Problems

Using the system configuration in Fig. 4.3 and the lower LFT, the $\mathcal{H}_\infty$ optimal control problem can be formulated as

($\mathcal{H}_\infty$ **optimal control problem**) "*Find the controller such that* $\mathcal{F}_l(M, K)$ *is internally stable and* $||\mathcal{F}_l(M, K)||_\infty$ *is minimized.*"

In practice, instead of the optimal control problem, the following $\mathcal{H}_\infty$ suboptimal control problem may be considered:

($\mathcal{H}_\infty$ **suboptimal control problem**) " *For given* $\gamma > 0$, *find a controller such that* $\mathcal{F}_l(M, K)$ *is internally stable and* $||\mathcal{F}_l(M, K)||_\infty < \gamma$."

Moreover, we introduce a class of problems, called the "$\mathcal{H}_\infty$ control standard problem." As the name suggests, this class of problems include a lot of general control problems in practices, further it is well known that even other types of problems can be reduced to this problem with some modifications. In fact, the problem for OBM control falls into this category. For the standard problems, the solution techniques have been well developed and the corresponding software products are available, for instance, MATLAB®, *Robust Control Toolbox*.

Consider the state-space formulation to represent the generalized plant M in Fig. 4.3 as follows:

$$\dot{\boldsymbol{x}} = A\boldsymbol{x} + B_1\boldsymbol{w} + B_2\boldsymbol{u}, \tag{4.5a}$$

$$\boldsymbol{z} = C_1\boldsymbol{x} + D_{11}\boldsymbol{w} + D_{12}\boldsymbol{u}, \tag{4.5b}$$

$$\boldsymbol{y} = C_2\boldsymbol{x} + D_{21}\boldsymbol{w} + D_{22}\boldsymbol{u}, \tag{4.5c}$$

where $\boldsymbol{x}$ is the state, $\boldsymbol{u}$ is the control input, $\boldsymbol{y}$ is the output and fed back to the controller, $\boldsymbol{w}$ and $\boldsymbol{z}$ are the fictitious input and output for evaluation in the sense of $\mathcal{H}_\infty$ norm, with compatible dimensions respectively.

Then, we assume the following:

(A1) (A, B_2) is stabilizable and (C_2, A) is detectable.
(A2) D_{12} is full column rank and D_{21} is full row rank.
(A3) $\begin{bmatrix} A - i\omega I & B_2 \\ C_1 & D_{12} \end{bmatrix}$ has full column rank for $\forall \omega \in \mathbb{R}$.
(A4) $\begin{bmatrix} A - i\omega I & B_1 \\ C_2 & D_{21} \end{bmatrix}$ has full row rank for $\forall \omega \in \mathbb{R}$.

The solvability condition for the problem and the specific formulation of the resultant controller are not presented in this monograph, which are not necessary to understand the essential ideas and the story in this monograph. But interested readers can refer to, e.g., [16, 28, 30, 115] for details. There exist two typical and popular approaches to solve the $\mathcal{H}_\infty$ control problems, one of which is the one to solve two algebraic Riccati equations, so-called "γ-iteration" [16, 28], and the other one is based on linear matrix inequalities (LMIs) [44, 80]. Both the algorithms can be available by using MATLAB®, *Robust Control Toolbox*.

Remark 4.1 If $\mathcal{F}_l(M, K)$ is interconnected with the uncertainty model Δ in the feedback configuration, then the $\mathcal{H}_\infty$ control problem becomes the robust control problem in the presence of Δ. Further, if Δ is stable, that is, $\Delta \in \mathcal{H}_\infty$ and $||\Delta||_\infty < 1/\gamma$, then according to Theorem 4.1 (small-gain theorem) the stabilizing controller K such that $||\mathcal{F}_l(M, K)||_\infty \leq \gamma$ ensures the robust stability of the entire control system, which is summarized in the following theorem:

Theorem 4.2 (Robust stability for unstructured uncertainties) *Let the feedback control system $\mathcal{F}_l(M, K)$ be stable and be interconnected with the stable uncertainty system $\Delta(s) \in \mathcal{H}_\infty$ bounded as $||\Delta(s)||_\infty < 1/\gamma$ $(\gamma > 0)$ in the feedback configuration. The entire closed-loop system is robustly stable, if and only if $||\mathcal{F}_l(M, K)||_\infty \leq \gamma$.*

4.2.3 Structured Uncertainties and μ-Analysis and Synthesis

If the uncertainty model matrix Δ is not given any specific structure, then it is called a full matrix and an "unstructured uncertainty". Otherwise, Δ is called a "structured uncertainty". In the case of structured uncertainties, the controller K mentioned in Remark 4.1 can be conservative. "The structured singular value" theory [4, 17, 72, 92] was developed to accommodate such cases in both robustness analysis and control design.

Without loss of generality, the structure of square uncertainty $\Delta \in \mathbf{C}^{n\times n}$ is given by the set as

$$\mathbf{\Delta} = \left\{\text{blockdiag}[\delta_1 I_{r1}, \ldots, \delta_s I_{rs}, \Delta_1, \ldots, \Delta_f] : \delta_i \in \mathbf{C}, \Delta_j \in \mathbf{C}^{m_j\times m_j}\right\}, \quad (4.6)$$

where $\sum_{i=1}^{s} r_i + \sum_{j=1}^{f} m_j = n$. Then, the structured singular value with respect to $\mathbf{\Delta}$ is first defined for a constant matrix $M \in \mathbf{C}^{n\times n}$ as in the following.

$$\mu_{\mathbf{\Delta}}(M) := \frac{1}{\min\{\bar{\sigma}(\Delta) : \Delta \in \mathbf{\Delta}, \det(I - M\Delta) = 0\}}, \quad (4.7)$$

where $\bar{\sigma}(\cdot)$ denotes the largest singular value. If there exists no $\Delta \in \mathbf{\Delta}$ which makes the matrix singular, then $\mu_{\mathbf{\Delta}}(M) := 0$. Furthermore, this notion can be extended to a linear time-invariant system matrix $M(s) \in \mathbf{C}^{n\times n}$ as

$$\mu_{\mathbf{\Delta}}(M(s)) := \sup_{\omega\in\mathbb{R}} \mu_{\mathbf{\Delta}}(M(i\omega)). \quad (4.8)$$

The method to analyze robustness of the system using μ instead of $||M(s)||_\infty = \sup_{\omega\in\mathbb{R}}(\bar{\sigma}/M(i\omega))$ (when $M(s) \in \mathcal{H}_\infty$) is called "$\mu$-analysis". As described above, when evaluating the robust stability of the feedback configuration of the structured Δ and $\mathcal{F}_l(M, K)$, the method using only their $\mathcal{H}_\infty$ norms will provide a conservative result. Hence, the μ-analysis is a less conservative approach in such a case, and the counterpart of Theorem 4.2 for structured uncertainties is expressed as in the following:

Theorem 4.3 (Robust stability for structured uncertainties) *Let the feedback control system $\mathcal{F}_l(M, K)$ be stable and be interconnected with the stable uncertainty system $\Delta(s) \in \mathcal{H}_\infty$ bounded as $||\Delta(s)||_\infty < 1/\gamma$ $(\gamma > 0)$, which has a structure that $\Delta(i\omega) \in \mathbf{\Delta}$ $for \forall\omega \in \mathbb{R}$, in the feedback configuration. The entire closed-loop system is robustly stable, if and only if $\mu_{\mathbf{\Delta}}(\mathcal{F}_l(M, K)) \leq \gamma$.*

In general, μ cannot be directly computed. The alternative method to compute μ is based on its upper and lower bounds. In particular, in practical applications, the upper bound is utilized because it can provide the sufficient condition for robustness. Therefore, here we introduce only the upper bound. Consider the following matrix set where a matrix has a compatible dimension with $\Delta \in \mathbf{\Delta}$:

$$\mathbf{D_\Delta} = \left\{\text{blockdiag}[D_1, \ldots, D_s, d_1 I_{m1}, \ldots, d_f I_{mf}] : D_i \in \mathbf{C}^{r_i \times r_i}, D_i = D_i^* \succ 0, d_j > 0 \in \mathbb{R}\right\}. \quad (4.9)$$

$D \in \mathbf{D_\Delta}$ is a scaling matrix compatible with $\Delta \in \mathbf{\Delta}$. Then, the following inequality always holds for a constant matrix M:

$$\mu_{\mathbf{\Delta}}(M) \leq \inf_{D \in \mathbf{D_\Delta}} \bar{\sigma}(DMD^{-1}). \quad (4.10)$$

Next, we address a control design method based on μ called "μ-synthesis", which utilizes the μ upper bound in (4.10). In the μ-synthesis, considering Theorem 4.3, the objective is to synthesize stabilizing a controller K and frequency-dependent scalings $D(i\omega)$'s $\in \mathbf{D_\Delta}$ such that

$$\sup_{\omega \in \mathbb{R}} \inf_{D \in \mathbf{D_\Delta}} \bar{\sigma}\left[D\mathcal{F}_l(M, K)D^{-1}(i\omega)\right] < \gamma. \quad (4.11)$$

This searching procedure is to iteratively repeat steps, and each step which consists of two sub-steps such that K or D is alternately searched while keeping the other one fixed. Therefore, this procedure is called the "D–K iteration."

Remark 4.2 Note that the above μ-based approaches are available in theory for time-invariant uncertainties. As described later, in the control problem of OBMs time-varying uncertainties need to be dealt with. In the case of time-varying uncertainties, the frequency-dependent scalings $D(i\omega)$'s have to be replaced by a constant scaling over $\forall\omega \in \mathbb{R}$ in both the μ-analysis and synthesis [3, 43, 68, 76, 79, 86, 93]. However, in general, the D–K iteration using a constant D is difficult to find an acceptable solution. Therefore, in this monograph, we will demonstrate the control design using the conventional D–K iteration and the μ-analysis with a constant scaling. Many research efforts have been devoted to the constant D–K iteration, e.g., [3, 79, 109].

4.2.4 TDOF Control System Structure

It is well known that $\mathcal{H}_\infty$ control is generally a very powerful tool to achieve frequency domain objectives, however, it has a shortcoming in that one cannot straightforwardly incorporate time domain objectives, such as overshoots, by using it alone. Therefore, when necessary, a two-degree-of-freedom (TDOF) control system structure [10, 30, 75] is utilized together with the $\mathcal{H}_\infty$ control. See Fig. 4.4 which shows a typical TDOF control system structure. P_{mdl} represents the model function which has a desirable time domain property, and the control system will try to let the closed-loop system exactly match P_{mdl} [75].

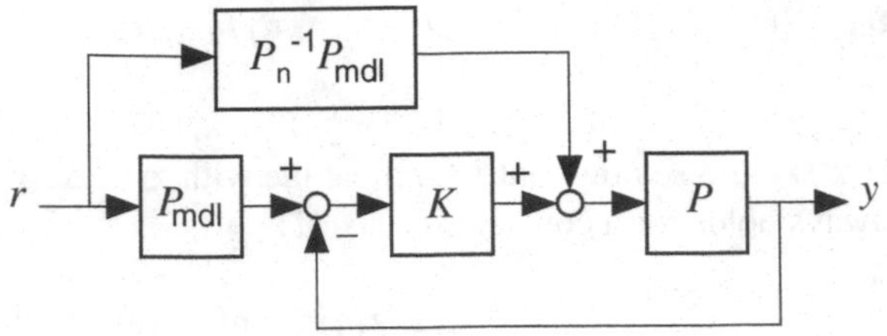

Fig. 4.4 TDOF control system structure; P actual plant, P_n nominal plant, P_{mdl} model plant, K controller, y output to be controlled, r reference command

4.3 Extended Matrix Polytopes for Model Uncertainty Representation

4.3.1 Model Uncertainties and Extended Matrix Polytopes

In many applications of offshore mechanical systems, the payload variation is the general case, which leads to the inertia parameter variation of the system as demonstrated in Chap. 3, e.g., when dealing with mineral resources, fishery resources, etc. In this section, we address the way of accommodating such model uncertainties for $\mathcal{H}_\infty$ control design and μ synthesis. The control design method via $\mathcal{H}_\infty$ control presented in this chapter requires a special treatment to represent model uncertainties in the inertia matrix. Hence, the objective here is to represent the uncertainties in an LFT form so that the problem can be dealt with in the $\mathcal{H}_\infty$ control framework. The aim of the following argument is to represent model uncertainties in a compatible manner with LFTs, as less conservatively as possible, depending on given information about those uncertainties. First, the motivation is addressed and the proposed methodology is outlined.

Suppose a matrix $X \in \mathbb{R}^{2\times 2}$ with two perturbational parameters Δx_{11} and Δx_{22} as described by the following:

$$X = \begin{bmatrix} x_{11} & x_{12} \\ x_{21} & x_{22} \end{bmatrix} = X_n + \Delta X,$$

$$X_n = \begin{bmatrix} x_{11n} & x_{12n} \\ x_{21n} & x_{22n} \end{bmatrix}, \quad \Delta X = \mathrm{diag}[\Delta x_{11}, \Delta x_{22}], \tag{4.12}$$

where X_n and ΔX denote the nominal matrix and the perturbational matrix which represents uncertainties of X respectively. When one knows only the bounds of those perturbations, $|\Delta x_{11}| \le \beta_1$ and $|\Delta x_{22}| \le \beta_2$, the possible perturbational parameter set is depicted by the solid lined Area 1 in Fig. 4.5, which shows the Δx_{11}–Δx_{22} space. When designing robust controllers for those perturbations, one must consider all the perturbations in Area 1. However, with more information about the dependence of those, e.g., that perturbations only occur with a common sign, which implies that all perturbed parameters simultaneously increase or decrease, the possible parameter set can be restricted to a smaller area such as the dashed lined Area 2. When assuming

Fig. 4.5 Various ranges on the Δx_{11}–Δx_{22} plane

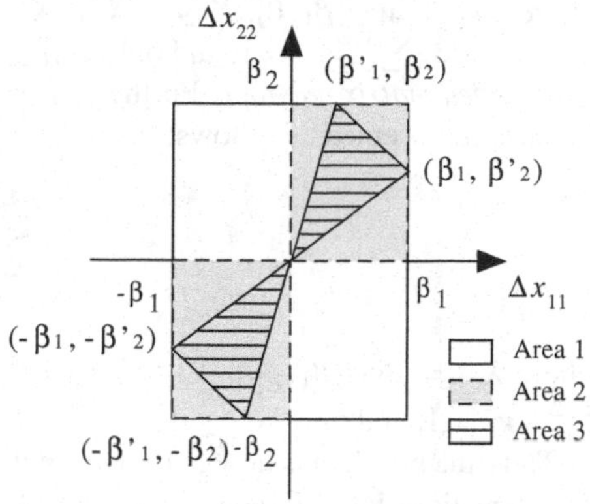

the inertia matrix perturbations due to payload variation, in fact, which is often the case, which has been already analyzed for the illustrative model in Chap. 3. In accordance with the increase of the payload for the manipulator, all or some entries in $M(\boldsymbol{q})$ in (2.22) will increase and vice versa. Then it is obviously conservative to consider all of Area 1 for such a case. Further with more detailed information, such as dependency between the perturbational parameters, one can take more effective design approaches as depicted by Area 3 in Fig. 4.5.

Which approach to be taken depends on the situations. In particular, when one is dealing with payload variation cases, the approach associated with Area 3 can possibly be taken. In the sequel, an appropriate method to represent the perturbation space in an LFT is introduced. Then, in Sect. 4.4, by using the numerical example of the illustrative model in Chap. 3, this notion will be demonstrated in detail.

One conventional tool to represent convex hulls such as Area 1, the *matrix polytope* [31, 105], can be available. For example, a representation of Area 1 using a matrix polytope is as in the following:

$$\Delta X = \sum_{i=1}^{4} \lambda_i X_{ai}, \tag{4.13}$$

where $X_{a1} = \mathrm{diag}[\beta_1, \beta_2]$, $X_{a2} = \mathrm{diag}[-\beta_1, \beta_2]$, $X_{a3} = -X_{a1}$, $X_{a4} = -X_{a2}$, $\lambda_i \in \mathbb{R}$, $0 \leq \lambda_i \leq 1$ and $\sum_{i=1}^{4} \lambda_i \leq 1$. X_{ai}'s correspond to the respective vertices of Area 1 and are therefore called *vertex matrices*.

Then, let us consider how to represent Areas 2 and 3. As seen from the figure, these are not convex; hence matrix polytopes cannot be directly applied. However, by introducing scaling parameters, this polytope approach can be easily extended so as to accommodate cases such as Area 2 as follows:

$$\Delta X = \nu_4 \sum_{i=1}^{3} \nu_i X_{bi}, \tag{4.14}$$

where $X_{b1} = \text{diag}[\beta_1, 0]$, $X_{b2} = X_{a1}$, $X_{b3} = \text{diag}[0, \beta_2]$, $\nu_i \in \mathbb{R}$, $0 \leq \nu_i \leq 1$ for $i = 1, \ldots, 3$, $\sum_{i=1}^{3} \nu_i = 1$, and only $|\nu_4| \leq 1$, which representation is referred to as an *extended matrix polytope*. Further, by using a completely identical manner, Area 3 is also represented as follows:

$$\Delta X = \kappa_3 \sum_{i=1}^{2} \kappa_i X_{ci}, \tag{4.15}$$

where $X_{c1} = \text{diag}[\beta_1, \beta_2']$, $X_{c2} = \text{diag}[\beta_1', \beta_2]$, $\kappa_i \in \mathbb{R}$, $0 \leq \kappa_i \leq 1$ for $i = 1, 2$, $\sum_{i=1}^{2} \kappa_i = 1$, and only $|\kappa_3| \leq 1$.

Thus, the problem is how to transform the above polytope and extended polytope representations into LFTs, which is, in fact, possible. One systematic transformation procedure for general cases is derived in the next section, and with which polytopes and extended polytopes can be dealt with in a uniform manner.

4.3.2 LFT Representation of an Extended Matrix Polytope

In order to design and analyze control systems with model uncertainties in the $\mathcal{H}_\infty$ control framework, it is required and powerful to represent the uncertainties in LFT forms. To this end, here an LFT representation of an extended matrix polytope is derived as in the following procedure: the uncertainty parameter set of an extended matrix polytope, which has an element-by-element dependent constraint such as $\sum_{i=1}^{3} \nu_i = 1$ in (4.14), is transformed into a parameter set with independent constraints; for this parameter set an LFT is derived; then the range of this parameter set is normalized by using another LFT; finally an LFT of the extended matrix polytope with the normalized uncertainty parameter set is derived by combining the above two LFTs as a Redheffer star product [17, 115].

Theorem 4.4 *Suppose a real matrix described by an extended matrix polytope as*

$$\Delta X = \nu_{l+1} \sum_{i=1}^{l} \nu_i X_i \in \mathbb{R}^{n \times m} \tag{4.16}$$

where l is a finite natural number with $1 \leq l < \infty$, X_i's are constant vertex matrices, uncertainty parameters ν_i's have dependent constraints that $0 \leq \nu_i \leq 1$ and $\sum_{i=1}^{l} \nu_i = 1$ for $i = 1, \ldots, l$, and that $|\nu_{l+1}| \leq 1$. Then, the extended matrix-polytope-based model can be represented in an LFT form as follows:

$$\Delta X = \mathcal{F}_u(\mathcal{S}(\Psi_2, \Psi_1), \Delta) \tag{4.17}$$

$$\Psi_1 = \left[\begin{array}{ccccc|c} \mathbf{0} & \mathbf{0} & \mathbf{0} & \dots & \mathbf{0} & X_1 - X_2 \\ I_n & \mathbf{0} & \mathbf{0} & \dots & \mathbf{0} & X_2 - X_3 \\ \mathbf{0} & I_n & \mathbf{0} & \dots & \mathbf{0} & X_3 - X_4 \\ \vdots & \ddots & \ddots & \ddots & \vdots & \vdots \\ \mathbf{0} & \dots & \mathbf{0} & I_n & \mathbf{0} & X_l \\ \hline \mathbf{0} & \dots & \mathbf{0} & \mathbf{0} & I_n & \mathbf{0} \end{array}\right] \in \mathbb{R}^{n(l+1)\times(nl+m)} \tag{4.18}$$

$$\Psi_2 = \left[\begin{array}{cc|cc} \multicolumn{2}{c|}{\mathbf{0}} & \multicolumn{2}{c}{I_{nl}} \\ \hline \frac{1}{2}I_{n(l-1)} & \mathbf{0} & \frac{1}{2}I_{n(l-1)} & \mathbf{0} \\ \mathbf{0} & I_n & \mathbf{0} & \mathbf{0} \end{array}\right] \in \mathbb{R}^{2nl\times 2nl} \tag{4.19}$$

$$\Delta = \text{blockdiag}[\delta_1 I_n, \dots, \delta_l I_n] \tag{4.20}$$

where δ_i's are normalized real perturbational parameters with $|\delta_i| \le 1$, and $\mathcal{S}(\cdot,\cdot)$ denotes a Redheffer star product.

Proof First, we prove that this representation with the dependent constraints in (4.16) is transformed into the following with the independent constraints with respect to the uncertainty parameter ξ_i's:

$$\Delta X = \xi_l \sum_{k=1}^{l} \{(1-\xi_{k-1}) \prod_{j=k}^{l-1} \xi_j\} X_k \tag{4.21}$$

where ξ's have independent constraints that $0 \le \xi_j \le 1$ for $j = 1, \dots, l-1$, $\xi_0 = 0$, and $\xi_l = \nu_{l+1}$. To prove that the representations (4.16) and (4.21) are equivalent, it suffices to prove that the following l-tuple parameter sets $\boldsymbol{\nu}$ and $\boldsymbol{\rho}$ coincide.

$$\boldsymbol{\nu} := \{(\nu_1, \dots, \nu_l) : 0 \le \nu_i \le 1, \sum_{i=1}^{l} \nu_i = 1\}$$

$$\boldsymbol{\rho} := \left\{ \begin{array}{l} (\rho_1, \dots, \rho_l) : \rho_i = (1-\xi_{i-1}) \prod_{j=i}^{l-1} \xi_j,\ 0 \le \xi_i \le 1 \\ \text{for } i = 1, \dots, l-1,\ \xi_0 = 0 \end{array} \right\} \tag{4.22}$$

By applying some algebraic manipulations to $\boldsymbol{\rho}$, it is immediately obtained that $\sum_{i=1}^{l} \rho_i = 1$ and $0 \le \rho_i \le 1$; thus $\boldsymbol{\rho} \subset \boldsymbol{\nu}$. Then, we consider the converse. Let $\boldsymbol{\nu}_0 = (\nu_{01}, \dots, \nu_{0l}) \in \boldsymbol{\nu}$ with $0 \le \nu_{0i} < 1$, and then there exist the unique ξ_{0i}'s such that $\nu_{0i} = (1-\xi_{0(i-1)}) \prod_{j=i}^{l-1} \xi_{0j}$, which is obtained in the following recursive manner:

$$
\begin{aligned}
\xi_{0(l-1)} &= 1 - \nu_{0l} \\
\xi_{0(l-2)} &= 1 - \frac{1}{\xi_{0(l-1)}}\nu_{0(l-1)} \\
&\vdots \\
\xi_{01} &= 1 - \frac{1}{\xi_{0(l-1)}\xi_{0(l-2)}\cdots\xi_{02}}\nu_{02}
\end{aligned}
$$

and $0 < \xi_{0i} \le 1$ is easily confirmed. Next, let us consider the case where $\nu_{0s} = 1\,(1 \le s \le l)$ and $\nu_{0i} = 0\,(i \ne s)$. In this case, by setting $\xi_{0(s-1)} = 0$ and $\xi_{0i} = 1\,(i \ne s-1, 0)$, ν_{0i} can be represented by the same form $\nu_{0i} = (1-\xi_{0(i-1)})\prod_{j=i}^{l-1}\xi_{0j}$. Hence, $\boldsymbol{\nu} = \boldsymbol{\rho}$ is proved.

Then, we can proceed to an LFT representation of the extended matrix polytope. In fact, (4.21) can be described by the following LFT:

$$
\begin{aligned}
\Delta X &= \mathcal{F}_u(\Psi_1, \Xi) \\
\Psi_1 &= \left[\begin{array}{ccccc|c}
\mathbf{0} & \mathbf{0} & \mathbf{0} & \ldots & \mathbf{0} & X_1 - X_2 \\
I_n & \mathbf{0} & \mathbf{0} & \ldots & \mathbf{0} & X_2 - X_3 \\
\mathbf{0} & I_n & \mathbf{0} & \ldots & \mathbf{0} & X_3 - X_4 \\
\vdots & \ddots & \ddots & \ddots & \vdots & \vdots \\
\mathbf{0} & \ldots & \mathbf{0} & I_n & \mathbf{0} & X_l \\
\hline
\mathbf{0} & \ldots & \mathbf{0} & \mathbf{0} & I_n & \mathbf{0}
\end{array}\right] \in \mathbb{R}^{n(l+1)\times(nl+m)} \\
\Xi &= \mathrm{blockdiag}[\xi_1 I_n, \ldots, \xi_l I_n],
\end{aligned}
\tag{4.23}
$$

where I denotes the identity matrix (when it is necessary to show the dimension n explicitly n will be added as the subscript) and $\mathbf{0}$ denotes a zero matrix with the appropriate dimension.

Further, the perturbational matrix Ξ is transformed into the LFT with a normalized perturbational matrix as follows:

$$
\begin{aligned}
\Xi &= \mathcal{F}_u(\Psi_2, \Delta) \\
\Psi_2 &= \left[\begin{array}{c|c}
\mathbf{0} & I_{nl} \\
\hline
\begin{array}{cc} \frac{1}{2}I_{n(l-1)} & \mathbf{0} \\ \mathbf{0} & I_n \end{array} & \begin{array}{cc} \frac{1}{2}I_{n(l-1)} & \mathbf{0} \\ \mathbf{0} & \mathbf{0} \end{array}
\end{array}\right] \in \mathbb{R}^{2nl\times 2nl} \\
\Delta &= \mathrm{blockdiag}[\delta_1 I_n, \ldots, \delta_l I_n]
\end{aligned}
\tag{4.24}
$$

where δ_i's are normalized real perturbational parameters with $|\delta_i| \le 1$.

Consequently, by combining the above two LFTs in (4.23) and (4.20) as the star product, the final formula is obtained:

$$
\begin{aligned}
\Delta X &= \mathcal{F}_u(\Psi_1, \mathcal{F}_u(\Psi_2, \Delta)) \\
&= \mathcal{F}_u(\mathcal{S}(\Psi_2, \Psi_1), \Delta).
\end{aligned}
\tag{4.25}
$$

Thus, the proof is completed.

Furthermore, Theorem 4.4 can be also applied to represent a conventional matrix polytope in an LFT form as presented in the following corollary:

Corollary 4.1 *Suppose a real matrix described by a conventional matrix polytope as*

$$\Delta X = \sum_{i=1}^{l} \lambda_i X_i \in \mathbb{R}^{n\times m} \tag{4.26}$$

where l is the number of the constant vertex matrices X_i's, $0 \le \lambda_i \le 1$ $(i = 1, \ldots, l)$ and $\sum_{i=1}^{l} \lambda_i \le 1$. Then, the matrix polytope can be represented in an LFT form as follows:

$$\Delta X = \mathcal{F}_u(\mathcal{S}(\Psi_2', \Psi_1), \Delta), \tag{4.27}$$

$$\Psi_2' = \left[\begin{array}{c|c} \mathbf{0} & I_{nl} \\ \hline \frac{1}{2} I_{nl} & \frac{1}{2} I_{nl} \end{array}\right] \in \mathbb{R}^{2nl\times 2nl}, \tag{4.28}$$

$$\Delta = \text{blockdiag}[\delta_1 I_n, \ldots, \delta_l I_n], \tag{4.29}$$

where Ψ_1 is the same as in (4.18), δ_i's are normalized real perturbational parameters with $|\delta_i| \le 1$.

Proof First, ΔX in (4.26) can be rewritten as:

$$\Delta X = \nu_{l+1} \sum_{i=1}^{l} \nu_i X_i \in \mathbb{R}^{n\times m}, \tag{4.30}$$

where $0 \le \nu_i \le 1$, $\sum_{i=1}^{l} \nu_i = 1$ for $i = 1, \ldots, l+1$. By noting that the range of ν_{l+1} is not $[-1, 1]$ but $[0, 1]$ as ν_i $(i \ne l+1)$, Theorem 4.4 can be applied with a modification of Ψ_2 as:

$$\Psi_2' = \left[\begin{array}{c|c} \mathbf{0} & I_{nl} \\ \hline \frac{1}{2} I_{nl} & \frac{1}{2} I_{nl} \end{array}\right] \in \mathbb{R}^{2nl\times 2nl}, \tag{4.31}$$

which yields (4.27). Thus, the proof is completed.

4.4 Control Design and Analysis

This section presents our proposed control design method exploiting $\mathcal{H}_\infty$ control to solve the problem defined in Chap. 2. The control design procedure consists of several parts, which are introduced in the sequel as follows. First, a nonlinear state-feedback scheme to reduce the nonlinearity of the system is introduced and then the resulting system is defined as a *virtual linear plant*. Second, the frequency-dependent control objectives and a sensitivity function shaping strategy using linear state-feedback control to achieve those objectives are presented, which is central to dealing with disturbances due to the base oscillation. Next, the uncertainty representation via an extended matrix polytope for the inertia matrix is discussed with some detail using the illustrative model and physical parameters in Table 3.4. After the preceding procedures, the problem is finally reduced to the framework of $\mathcal{H}_\infty$ control.

4.4.1 Nonlinear State Feedback and Virtual Linear Plant

To reduce the nonlinearity of the dynamical model of OBMs in (2.1) except the base oscillation induced disturbance $H(\boldsymbol{q}, \dot{\boldsymbol{q}}, \dot{\boldsymbol{q}}_{\rm b}, \ddot{\boldsymbol{q}}_{\rm b})$ and to construct a *virtual linear plant*, the following typical nonlinear state-feedback scheme based on the nominal parameters is employed.

$$\boldsymbol{\tau} = C_n(\boldsymbol{q}, \dot{\boldsymbol{q}})\dot{\boldsymbol{q}} + G_n(\boldsymbol{q}, \boldsymbol{q}_{\rm b}) + D\dot{\boldsymbol{q}} + M_n(\boldsymbol{q})\boldsymbol{u} \tag{4.32}$$

where the terms with the subscript $(\cdot)_n$ denote the corresponding terms with the nominal parameters in (2.1) as introduced in Sect. 3.3, and $\boldsymbol{u}$ is a new control torque vector.

Substituting (4.32) into the model (2.1) and then pre-multiplying by $M_n^{-1}(\boldsymbol{q})$ yield the following dynamical system, which is defined as a *virtual linear plant* with the time-varying perturbation of the inertia matrix $\Delta I(\boldsymbol{q})$ and the input-port disturbance $\boldsymbol{d}$.

$$(I + \Delta I(\boldsymbol{q}))\ddot{\boldsymbol{q}} = \boldsymbol{u} + \boldsymbol{d}, \tag{4.33}$$

where

$$\boldsymbol{d} = -M_n^{-1}(\boldsymbol{q})\{\Delta C(\boldsymbol{q}, \dot{\boldsymbol{q}})\dot{\boldsymbol{q}} + \Delta G(\boldsymbol{q}, \boldsymbol{q}_{\rm b}) + H(\boldsymbol{q}, \dot{\boldsymbol{q}}, \dot{\boldsymbol{q}}_{\rm b}, \ddot{\boldsymbol{q}}_{\rm b})\}, \tag{4.34}$$

$$\begin{aligned} \Delta I(\boldsymbol{q}) &= M_n^{-1}(\boldsymbol{q})(M(\boldsymbol{q}) - M_n(\boldsymbol{q})) \\ &= \begin{bmatrix} \Delta I_{11} \Delta I_{12} \\ \Delta I_{21} \Delta I_{22} \end{bmatrix}, \end{aligned} \tag{4.35}$$

$$\Delta C(\boldsymbol{q}, \dot{\boldsymbol{q}}) = C(\boldsymbol{q}, \dot{\boldsymbol{q}}) - C_n(\boldsymbol{q}, \dot{\boldsymbol{q}}), \tag{4.36}$$

$$\Delta G(\boldsymbol{q}, \boldsymbol{q}_{\rm b}) = G(\boldsymbol{q}, \boldsymbol{q}_{\rm b}) - G_n(\boldsymbol{q}, \boldsymbol{q}_{\rm b}). \tag{4.37}$$

The key points of the virtual linear plant in (4.33) are:

- in order to apply linear $\mathcal{H}_\infty$ control, the system is essentially regarded as a linear system;
- with respect to model uncertainties, the uncertainty of the inertia matrix, which is the most influential to the overall system, is explicitly modeled, while the other uncertainties in $C(\boldsymbol{q},\dot{\boldsymbol{q}})$ and $G(\boldsymbol{q},\boldsymbol{q}_\mathrm{b})$ are absorbed into the disturbance $\boldsymbol{d}$ together with the base-oscillation induced disturbance $H(\boldsymbol{q},\dot{\boldsymbol{q}},\dot{\boldsymbol{q}}_\mathrm{b},\ddot{\boldsymbol{q}}_\mathrm{b})$;
- excluding $\boldsymbol{d}$, the system from $\boldsymbol{u}$ to $\boldsymbol{q}$ are completely decoupled and consists of the same single-input single-output (SISO) systems, i.e., double integrators, in the nominal case.

It should be noted that the effects of $\Delta C(\boldsymbol{q},\dot{\boldsymbol{q}})$ and $\Delta G(\boldsymbol{q},\boldsymbol{q}_\mathrm{b})$ in $\boldsymbol{d}$ are much less than that of $H(\boldsymbol{q},\dot{\boldsymbol{q}},\dot{\boldsymbol{q}}_\mathrm{b},\ddot{\boldsymbol{q}}_\mathrm{b})$ and can be attenuated by applying linear state-feedback control as explained later, however, $H(\boldsymbol{q},\dot{\boldsymbol{q}},\dot{\boldsymbol{q}}_\mathrm{b},\ddot{\boldsymbol{q}}_\mathrm{b})$ is not the case and to be overcome by an $\mathcal{H}_\infty$ control scheme. Further, since the system can be considered to consist of the same SISO systems, the control design procedure in the sequel can be applied to each SISO system, which will make the control design method very simple regardless of the system order. Hence, this feature is one of the remarkable advantages of this method.

Here, we discuss the gravity compensator $G_n(\boldsymbol{q},\boldsymbol{q}_\mathrm{b})$ in (4.32). In the case of local coordinate problems, measurements of the base oscillation are not used to generate the reference trajectories to be tracked by the manipulator, and are used only for the gravity compensator $G_n(\boldsymbol{q},\boldsymbol{q}_\mathrm{b})$. Therefore, if the control system can work without the measurements, that is, any oscillation sensors are not necessary, the system can be constructed with lower cost, which will be practically advantageous.

Then, let us consider the gravity compensator without the information of $\boldsymbol{q}_\mathrm{b}$. For the illustrative model, its specific form is as follows:

$$
\begin{aligned}
G_\mathrm{l}(\boldsymbol{q}) &= [G_{\mathrm{l}1}, G_{\mathrm{l}2}]^T,\\
G_{\mathrm{l}1} &= -\{m_1a_1\sin(q_1) + m_{2n}(l_1\sin(q_1) + a_{2n}\sin(q_1+q_2))\}g,\\
G_{\mathrm{l}2} &= -m_{2n}a_{2n}\sin(q_1+q_2)g.
\end{aligned}
\tag{4.38}
$$

The subscript $(\cdot)_\mathrm{l}$ expresses that the compensator $G(\boldsymbol{q})$ is for local coordinate problems. Note that seemingly $G_\mathrm{l}(\boldsymbol{q})$ does not exactly match $G(\boldsymbol{q},\boldsymbol{q}_\mathrm{b})$ in (2.5), even in the nominal case, because of the lack of the measurement of q_b. Thus, to assess its effectiveness, the quantities defined by Eqs. (4.39) and (4.40) are examined, where the base oscillation q_b is represented by a single sinusoidal wave in (2.7).

$$
\alpha_1 := \frac{\omega_1}{2\pi}\int_0^{\frac{2\pi}{\omega_1}} \{\sin(A_1\sin(\omega_1 t)+q_1) - \sin(q_1)\}^2\,dt \tag{4.39}
$$

$$
\alpha_2 := \frac{\omega_1}{2\pi}\int_0^{\frac{2\pi}{\omega_1}} \sin^2(A_1\sin(\omega_1 t)+q_1)\,dt \tag{4.40}
$$

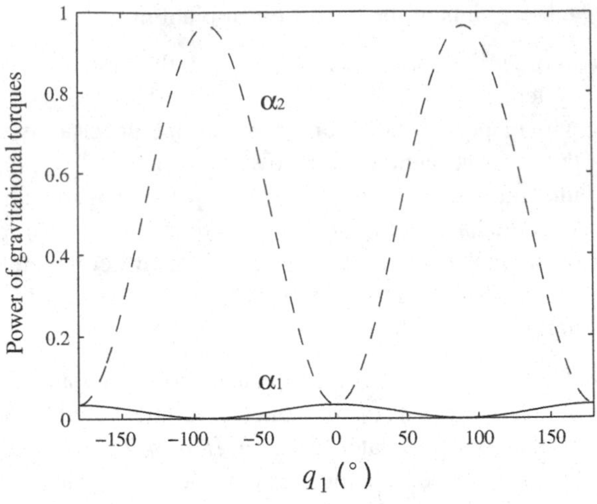

Fig. 4.6 Effectiveness of the gravitational compensator G_n in the case of $A_1 = 15°$; *solid line* α_1, *dashed line* α_2

where α_1 is associated with the power of the gravitational torque with the compensation, while α_2 without it. These values are invariant with respect to the base oscillation angular frequency ω_1 and can be interpreted as that if α_2 is lager than α_1, then it implies that the compensation scheme is effective. Then some numerical examples with $A_1 = 5$, 10 and 15° were examined for each q_1 and the case of 15° is shown in Fig. 4.6. As seen in the figure, α_2 is larger than α_1 over the range of q_1. The results of the other cases of 5 and 10° were similar to that and even better. Therefore, we have confirmed that the gravitational compensation can successfully suppress the gravitational torque. In fact, it can be demonstrated by simulations that the control system works effectively without measurements of the base oscillation for local coordinate problems.

4.4.2 Extended Matrix-Polytope-Based Model Uncertainty Representation for the Inertia Matrix

Here we present the way of representing model uncertainties in an LFT form for $\Delta I(\boldsymbol{q})$ in (4.35) of the virtual linear plant in (4.33). By employing the parameter variation data for the illustrative model shown in Table 3.4, we analyze how $\Delta I(\boldsymbol{q})$ varies according to the payload variation and $\boldsymbol{q}$. Then, the idea of extended matrix polytope in Sect. 4.3 will be applied to $\Delta I(\boldsymbol{q})$.

Recall the analysis on the parameter variation due to the payload change in Sect. 3.3. The inertia matrix variation $\Delta M(\boldsymbol{q}) = M(\boldsymbol{q}) - M_n(\boldsymbol{q})$ in (3.7) is linear in Δm_2. Thus, substituting (3.7) into (4.35) yields

$$\begin{aligned}\Delta I(\boldsymbol{q}) &= M_n^{-1}(\boldsymbol{q})\Delta M(\boldsymbol{q}) \\ &= \Delta m_2 M_n^{-1}(\boldsymbol{q})M'(\boldsymbol{q}),\end{aligned} \tag{4.41}$$

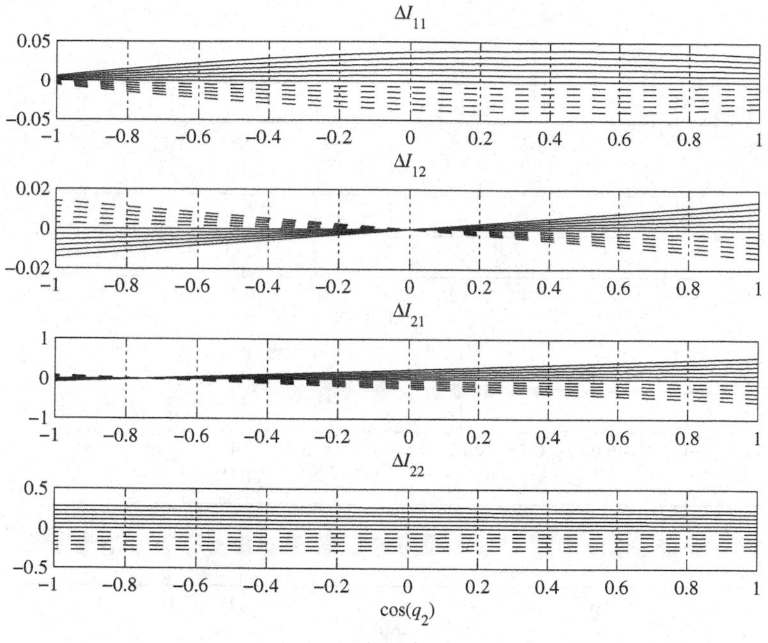

Fig. 4.7 ΔI_{ij}'s versus $\cos(q_2)$ with various payloads. The *solid lines* represents the cases of $\Delta m_2 > 0$, and the *dashed* one do the case of $\Delta m_2 < 0$

which shows that $\Delta I(\boldsymbol{q})$ is also linear in Δm_2. For all the parameter cases in Table 3.4, that is, the payload widths 0, 1, 2, . . ., 10 mm, each element $\Delta I_{ij}\,(i, j = 1, 2)$ in (4.41) is calculated over $q_2 \in [0, 180]°$, and the results of which are shown in Figs. 4.7, 4.8 and 4.9 respectively. Note that $\Delta I_{ij}\,(i, j = 1, 2)$ is an even function of q_2 and, thus, the calculation over nonnegative q_2 suffices.

Figure 4.7 shows that ΔI_{ij}'s versus $\cos(q_2)$ on the different scales on the respective vertical axes, where the solid lines represent the cases of positive variation of payload mass ($\Delta m_2 > 0$) and the dashed ones do the negative cases ($\Delta m_2 < 0$). As seen from the figure, ΔI_{ij}'s, except ΔI_{11}, are almost linearly dependent on $\cos(q_2)$. Then, see Fig. 4.8 which displays the same data as in Fig. 4.7 on a uniform vertical axis scale. It is seen that ΔI_{11} and ΔI_{12} are much less than ΔI_{21} and ΔI_{22}. If we regard the set of ΔI_{ij}'s as a set of four-dimensional vector in $\mathbb{R}^4$, the set can be a subset of two-dimensional manifold with coordinates $(\Delta m_2, \cos(q_2))$. Furthermore, the facts that ΔI_{ij}'s are completely linear in Δm_2 and almost linear in $\cos(q_2)$ considering that ΔI_{11} is much less than the other ΔI_{ij}'s suggest that we can approximate the set of ΔI_{ij} by a subset of a hyperplane in $\mathbb{R}^4$. To understand this claim intuitively, three-dimensional graphs of ΔI_{ij}'s are presented in Fig. 4.9, where the set is very similar to Area 3 in Fig. 4.5. Thus, the set of $\Delta I(\boldsymbol{q})$ can be represented by an extended matrix polytope with merely two vertex matrices as in (4.15) and subsequently by its LFT form. And see Table 4.1 that presents the minimum and maximum values of ΔI_{ij}. If one takes into account only the information in Table 4.1 to represent the model

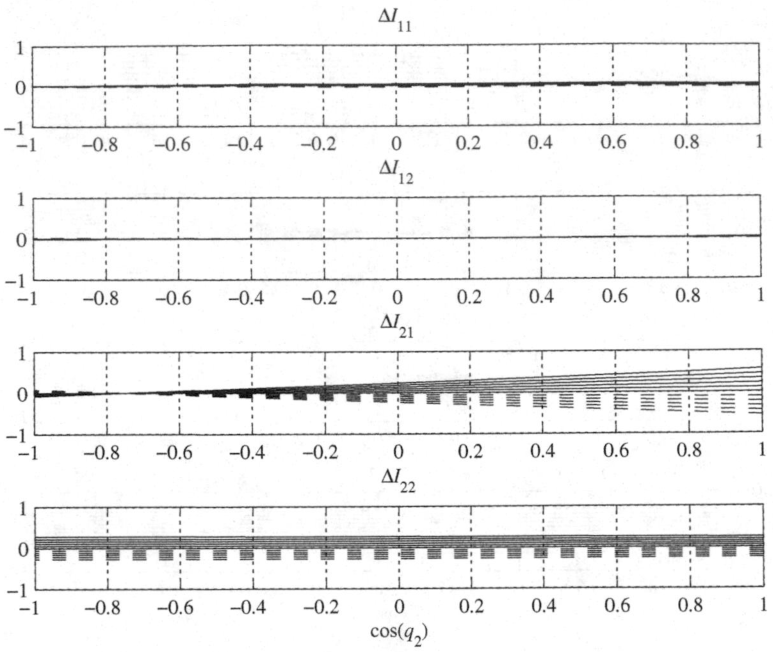

Fig. 4.8 Comparison of $\varDelta I_{ij}$'s on the uniform scale

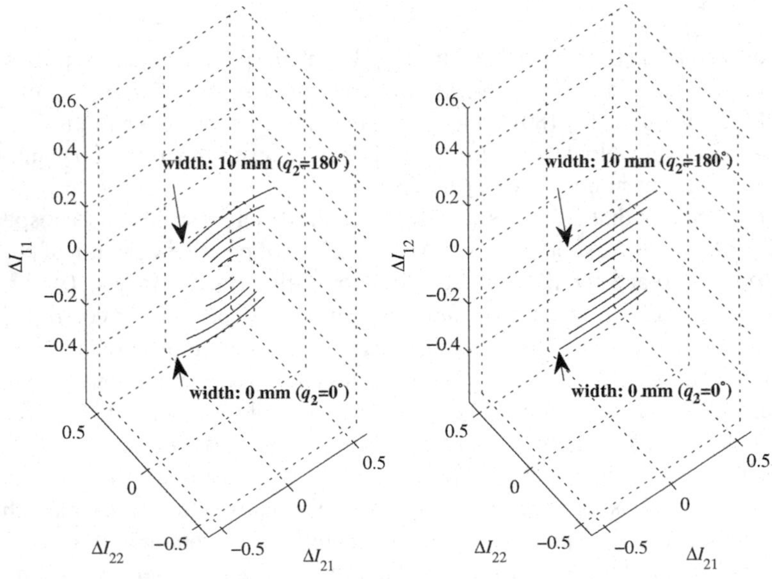

Fig. 4.9 3D representation of the $\varDelta I_{ij}$'s space

Table 4.1 Range of ΔI_{ij} variation

ΔI_{ij}	Min	Max
ΔI_{11}	−0.0393	0.0393
ΔI_{12}	−0.0141	0.0141
ΔI_{21}	−0.5591	0.5591
ΔI_{22}	−0.2740	0.2740

uncertainty, the representation is based on the conventional matrix polytope resulting a hyper-polyhedron with 16 vertices (recall Area 1 in Fig. 4.5), which is obviously too conservative. Therefore, The above argument justifies that the approach based on extended matrix polytopes is effective and significant.

Next, we analyze in a quantitative manner how the set of ΔI_{ij}'s (two-dimensional manifold) can be approximated by a hyperplane modeled with the following extended matrix polytope:

$$\Delta I(\boldsymbol{q}) \approx \nu_3(\nu_1 \Delta I_1 + \nu_2 \Delta I_2), \tag{4.42}$$

$$\Delta I_1 = \begin{bmatrix} 0.0336 & 0.0141 \\ 0.5591 & 0.2454 \end{bmatrix}, \tag{4.43}$$

$$\Delta I_2 = \begin{bmatrix} 0.0054 & -0.0141 \\ -0.0790 & 0.2736 \end{bmatrix}, \tag{4.44}$$

where ΔI_1 and ΔI_2 are vertex matrices respectively corresponding to the cases of $q_2 = 0°$ and $q_2 = 180°$ with the same payload width of 10 mm, $\nu_i \in \mathbb{R}$, $0 \le \nu_i \le 1$ for $i = 1, 2$, $\sum_{i=1}^{2} \nu_i = 1$, and $|\nu_3| \le 1$. ν_1 and ν_2 are associated with $\cos(q_2(t))$, and ν_3 with Δm_2. Thus, once the payload has been fixed, ν_3 is time-invariant, whereas $\nu_1$and ν_2 is time-varying during the manipulator motion.

In order to facilitate the analysis, let us consider a bijective mapping from $\mathbb{R}^{2\times 2}$ to $\mathbb{R}^4$ and its inverse defined as follows. Let $X \in \mathbb{R}^{2\times 2}$ be

$$X = \begin{bmatrix} X_{11} & X_{12} \\ X_{21} & X_{22} \end{bmatrix}, \tag{4.45}$$

then, the mapping $vec(\cdot)$ and its inverse $vec^{-1}(\cdot)$ are defined as:

$$vec(X) := [X_{11}, X_{12}, X_{21}, X_{22}]^T \in \mathbb{R}^4, \tag{4.46}$$

$$vec^{-1}(vec(X)) := X. \tag{4.47}$$

Hence, using the tools, we analyze sort of feasibility of the approximation of $\Delta I(\boldsymbol{q})$ in (4.41) by (4.42) quantitatively. Instead of the set of $\Delta I(\boldsymbol{q})$'s variation, we consider the set of $vec(\Delta I(\boldsymbol{q}))$' variation in $\mathbb{R}^4$ which is denoted by $\mathcal{M}$. Next, let $\mathcal{M}^*$ be the hyperplane spanned by $vec(\Delta I_1)$ and $vec(\Delta I_2)$. Let us define a orthogonal projection of $\boldsymbol{dI} \in \mathcal{M}$ onto $\mathcal{M}^*$ being denoted by $\boldsymbol{dI}^* = \mathcal{P}(\boldsymbol{dI}) \in \mathcal{M}^*$ (see

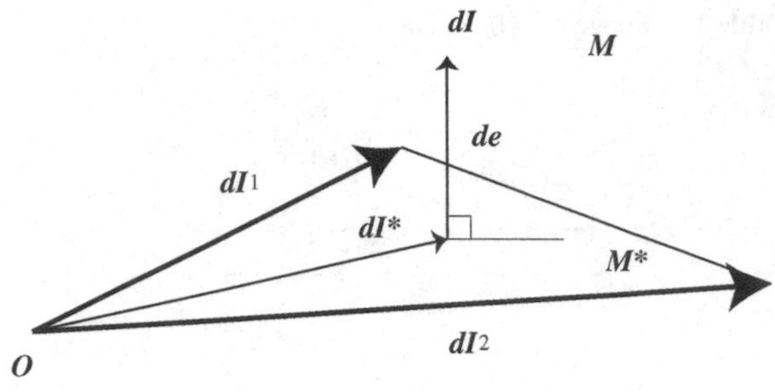

Fig. 4.10 Projection onto $\mathcal{M}^*$

Fig. 4.10). Since the approximation error vector $\boldsymbol{de} = \boldsymbol{dI} - \boldsymbol{dI}^*$ is orthogonal to $\mathcal{M}^*$, i.e., $vec(\Delta I_1)$ and $vec(\Delta I_2)$, $\boldsymbol{dI}^*$ for given $\boldsymbol{dI}$ can be obtained uniquely and straightforwardly as:

$$\boldsymbol{dI}^* = [\boldsymbol{dI}_1, \boldsymbol{dI}_2]\begin{bmatrix} <\boldsymbol{dI}_1 \cdot \boldsymbol{dI}_1> & <\boldsymbol{dI}_1 \cdot \boldsymbol{dI}_2> \\ <\boldsymbol{dI}_2 \cdot \boldsymbol{dI}_1> & <\boldsymbol{dI}_2 \cdot \boldsymbol{dI}_2> \end{bmatrix}^{-1}\begin{bmatrix} <\boldsymbol{dI}_1 \cdot \boldsymbol{dI}> \\ <\boldsymbol{dI}_2 \cdot \boldsymbol{dI}> \end{bmatrix}, \tag{4.48}$$

where $\boldsymbol{dI}_1 = vec(\Delta I_1)$, $\boldsymbol{dI}_2 = vec(\Delta I_2)$, and $< \cdot >$ denotes Euclidean inner product defined in $\mathbb{R}^4$.

Being based on (4.48), we calculate the following criterion to evaluate the approximation:

$$\varrho_1 = \frac{\max_{\boldsymbol{dI}\in\mathcal{M}} \|\boldsymbol{dI} - \boldsymbol{dI}^*\|}{\max_{\boldsymbol{dI}\in\mathcal{M}} \|\boldsymbol{dI}|}, \tag{4.49}$$

where $\|\cdot\|$ denotes the Euclidean norm. ϱ_1 indicates how close to a hyperplane geometrically $\mathcal{M}$ is in $\mathbb{R}^4$, that is, the smaller ϱ_1 implies the better approximation. In this case, the resulting ϱ was 0.0279, which suggests that the approximation is fairly good.

Moreover, using the inverse mapping $vec^{-1}(\cdot)$, we evaluate the feasibility of the approximation in $\mathbb{R}^{2\times 2}$. Specifically, the following criterion is computed:

$$\varrho_2 = \max_{\boldsymbol{dI}\in\mathcal{M}} \bar{\sigma}(vec^{-1}(\boldsymbol{dI} - \boldsymbol{dI}^*)), \tag{4.50}$$

where $\bar{\sigma}(\cdot)$ denotes the largest singular value. And the resulting ϱ_2 was 0.0170. $\Delta I(\boldsymbol{q})$ is a displacement matrix from the identity matrix I_2, hence the approximation error criterion ϱ_2 should be evaluated relatively to 1. In this sense, we can conclude that the approximation is fairly good.

Remark 4.3 The above argument is basically based on the viewpoint that how less conservative and simple model uncertainty representation can be for control design using the $\mathcal{H}_\infty$ control framework. For stringently evaluating the feasibility of the approximation, the robust stability of the resulting control system based on the uncertainty modeled by (4.42) should be analyzed over $vec^{-1}(\mathcal{M})$. However, at this point, the effectiveness and significance of the developed machinery, *extended matrix polytopes*, has been satisfactorily demonstrated.

Finally in this subsection, we derive an LFT form for $\Delta I(\boldsymbol{q})$ approximated by (4.42). Applying Theorem 4.4 to $\Delta I(\boldsymbol{q})$ in (4.42) yields the following LFT representation:

$$\Delta I(\boldsymbol{q}) = \mathcal{F}_u(\mathcal{S}(\Psi_2, \Psi_1), \Delta_I), \tag{4.51}$$

$$\Psi_1 = \left[\begin{array}{cc|c} \mathbf{0} & \mathbf{0} & \Delta I_1 - \Delta I_2 \\ I_2 & \mathbf{0} & \Delta I_2 \\ \hline \mathbf{0} & I_2 & \mathbf{0} \end{array}\right], \tag{4.52}$$

$$\Psi_2 = \left[\begin{array}{cc|cc} \multicolumn{2}{c|}{\mathbf{0}} & \multicolumn{2}{c}{I_4} \\ \hline \frac{1}{2}I_2 & \mathbf{0} & \frac{1}{2}I_2 & \mathbf{0} \\ \mathbf{0} & I_2 & \mathbf{0} & \mathbf{0} \end{array}\right], \tag{4.53}$$

$$\mathcal{S}(\Psi_2, \Psi_1) = \left[\begin{array}{cc|c} \mathbf{0} & \mathbf{0} & \Delta I_1 - \Delta I_2 \\ \frac{1}{2}I_2 & \mathbf{0} & \frac{1}{2}(\Delta I_1 + \Delta I_2) \\ \hline \mathbf{0} & I_2 & \mathbf{0} \end{array}\right], \tag{4.54}$$

$$\mathbf{\Delta}_I := \{\text{blockdiag}[\delta_1(t)I_2, \delta_2 I_2] : \delta_i \in \mathbb{R}, |\delta_i| \leq 1\}, \tag{4.55}$$

$$\Delta_I \in \mathbf{\Delta}_I, \tag{4.56}$$

where $\delta_1(t)$ is associated with $\cos(q_2(t))$, thus time-varying, whereas δ_2 is time-invariant correspondingly to Δm_2. Further, for the control design, the following equation will be utilized:

$$I + \Delta I(\boldsymbol{q}) = \mathcal{F}_u(\Phi, \Delta_I), \tag{4.57a}$$

$$\Phi = \begin{bmatrix} \Phi_{11} & \Phi_{12} \\ \Phi_{21} & \Phi_{22} \end{bmatrix} \tag{4.57b}$$

$$= \left[\begin{array}{cc|c} \mathbf{0} & \mathbf{0} & \Delta I_1 - \Delta I_2 \\ \frac{1}{2}I_2 & \mathbf{0} & \frac{1}{2}(\Delta I_1 + \Delta I_2) \\ \hline \mathbf{0} & I_2 & I_2 \end{array}\right]. \tag{4.57c}$$

Note that the above notion and methodology is obviously not exclusive for the OBM problems and extensively available for general robotic manipulator control problems.

4.4.3 *Sensitivity Function Shaping Strategy Using Linear State-Feedback Control*

In order to solve the oscillatory base problem, it is crucial to cope with disturbances due to the base oscillation, which is exerted at the input port of the plant. In this

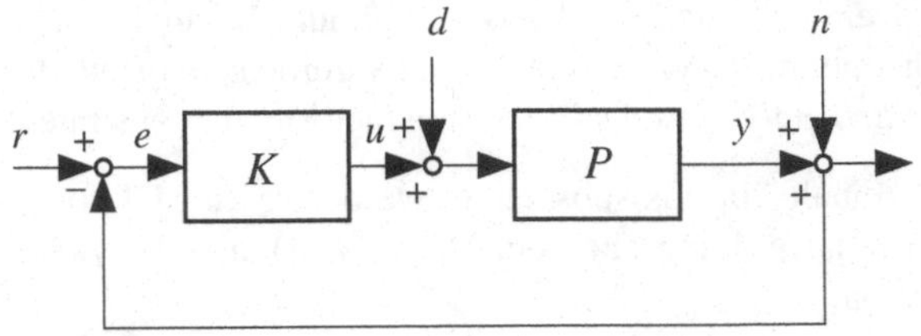

Fig. 4.11 Diagram of a closed-loop system

work, this issue is captured as a filtering problem, i.e., how to make the sensitivity of the system to the disturbance as low as possible. Hence, the issue is, specifically, to construct desirable frequency-dependent properties of the corresponding sensitivity functions of the system. In this section, we present a strategy for shaping sensitivity functions using linear state-feedback control for $\mathcal{H}_\infty$ control design.

By being based on the virtual linear plant in the nominal case, we consider each decoupled SISO system, that is, from u_i to q_i ($i = 1, 2$), whose state-space and output equations are as in the following:

$$\dot{\boldsymbol{x}} = \begin{bmatrix} 0 & 1 \\ 0 & 0 \end{bmatrix} \boldsymbol{x} + \begin{bmatrix} 0 \\ 1 \end{bmatrix} u_i, \tag{4.58a}$$

$$y_i = \begin{bmatrix} 1 & 0 \end{bmatrix} \boldsymbol{x}, \tag{4.58b}$$

where $\boldsymbol{x} = [q_i, \dot{q}_i]^T$. Then, Fig. 4.11 depicts a closed-loop system, where P and K represent the transfer functions of this SISO system and a controller to be designed respectively; r_i is the reference trajectory to be tracked by y_i; n is a sensor noise; $e_i = r_i - y_i - n$ is the tracking error with noise; and d_i is a disturbance at the input port. Then the following sensitivity functions are defined as standard [115].

$$S := (1 + PK)^{-1} \; (r_i \to e_i, n \to -e_i) \tag{4.59}$$

$$S_P := (1 + PK)^{-1} P \; (d_i \to -e_i) \tag{4.60}$$

$$T_a := K(1 + PK)^{-1} \; (r_i \to u_i, n \to -u_i) \tag{4.61}$$

S is referred to as *sensitivity function* which represents tracking control performance and noise sensitivity; pre-multiplying S by the plant P yields S_P, which pertains to the input port disturbance attenuation property and, hence, plays a key role in coping with disturbances due to the base oscillation; T_a is referred to as *quasi-complementary sensitivity function*. T_a is the sensitivity from r_i and n to the control input, and will be utilized not only when in need of suppressing the control input, but also when regarding an additive unmodeled dynamics. Furthermore, T_a is necessary to be taken into account so as to make the $\mathcal{H}_\infty$ control design problem the standard problem [16, 115].

Before introducing the method of shaping the above functions, the frequency domain characteristics of the disturbance $\boldsymbol{d}$ in (4.33) is evaluated. In particular, torque H in (2.6) is focused on, which will be dominant in $\boldsymbol{d}$. Assume the steady-state

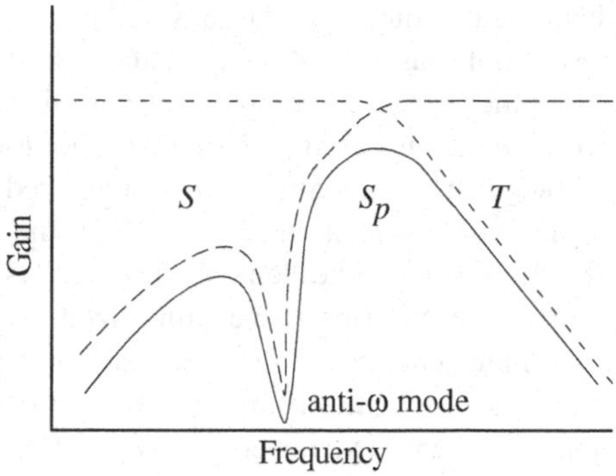

Fig. 4.12 Desirable frequency responses of the sensitivity functions

manipulator operating frequency ω_s and examine Eqs. (2.6) and (2.7), it is then found out that the principal frequency of H is expected to be ω_i and that $2\omega_i$, $\omega_s + \omega_i$ and $|\omega_s - \omega_i|$ will also appear since there exist the products of $\dot{q}_b$, $\dot{q}_1$, and $\dot{q}_2$. Then, the minimal frequency range Ω such that contains all those frequencies can be determined. Thus, the discussion is summarized as "to attenuate the effect of $\boldsymbol{d}$ it is necessary that S_P should be low enough over the frequency range Ω."

Now let us move on to the sensitivity function shaping strategy. Taking the above necessary S_P property into account together with the standard requirements for S and T, the desirable frequency responses of those functions are as depicted in Fig. 4.12. Focus on the anti-resonant modes of S and S_P, labeled with "anti-ω modes" in the figure, which will satisfy the above requirement for S_P. As seen from (4.59) and (4.60), once the plant is determined, S and S_P cannot be independently constructed by designing K. Therefore, the idea is to achieve this frequency-dependent objective for S_P by shaping S instead of S_P, which gives rise to anti-ω modes to S. It should be noted that the anti-ω modes for S is necessary for global coordinate problems where the reference trajectory necessarily contains ω_i's modes to cancel out the base oscillation while it is not the case for local coordinate ones. On the other hand, recall that the plant in (4.58) has a double integrator, which increases the gains of S_P at low frequencies as compared to those of S and then will degrade the disturbance attenuation performance. Therefore, to maintain the relation between S and S_P at low frequencies, a linear state-feedback control $F\boldsymbol{x} = [F_1, F_2]\boldsymbol{x}$ is added to the plant so that P has a flat gain property with less than zero dB at low frequencies and hence the requirements are achieved. Thus, the model in (4.58) is rewritten as in (4.62):

$$\dot{\boldsymbol{x}} = \begin{bmatrix} 0 & 1 \\ -F_1 & -F_2 \end{bmatrix} \boldsymbol{x} + \begin{bmatrix} 0 \\ 1 \end{bmatrix} u_i, \tag{4.62a}$$

$$y_i = \begin{bmatrix} 1 \; 0 \end{bmatrix} \boldsymbol{x}, \tag{4.62b}$$

where u_i is the new control input. F_1 which controls the relative gain of S_P to S should be chosen subject to the requirement for S_P, then F_2 can be determined such that with respect to the F_1 the nominal model in (4.62) has an allowable damping property.

Theoretically, the larger F_1 gain will lead to the smaller gain of S_p. However, in the presence of sensor error and actuator saturation, F_1 cannot be made arbitrarily large, and neither can F_2. Therefore, an optimization scheme is applied to determine the gain F, which is based on control simulations and numerical direct search.

For given F, a controller is synthesized and a step tracking control simulation is conducted as mentioned later. Then, the control error vector $\boldsymbol{e}$ is calculated. By extracting steady-state elements from $\boldsymbol{e}$, $\boldsymbol{e}_s$ is obtained, and its Frobenius norm $||\boldsymbol{e}_s||_F$ is set to the criterion to be minimized. Since the $\mathcal{H}_\infty$ control synthesis is via a suboptimal scheme and the obtained $||\boldsymbol{e}_s||_F$ is thus not continuous in F, nonlinear optimization schemes using the derivatives cannot be used. Instead, a direct search method was employed over $F_1 \times F_2 = [20, 21, \ldots, 89, 90] \times [20, 21, \ldots, 89, 90]$, which range had been determined in advance through a trial and error manner via simulations. The resultant F is [66, 24].

Consequently, the proposed strategy involving the linear state-feedback control scheme is reduced to a mixed-sensitivity problem for S and T_a for the plant in (4.62), where S_p will be indirectly shaped through shaping S respectively. Note that the linear state-feedback control scheme to shape S_P distinguishes our methodology from the conventional ones, e.g., [91], which employ a PD control to stabilize the double integrator after linearization.

4.4.4 Generalized Plant for $\mathcal{H}_\infty$ Control Design and Analysis

We discuss here construction of generalized plants for $\mathcal{H}_\infty$ control design and analysis. To demonstrate and evaluate the extended polytope-based model uncertainty representation and the weighting function matrix W_n for the complementary sensitivity function T_a, we consider four types of system configurations as depicted in Figs. 4.13 and 4.14, where P and K denote the two-input two-output plant and controller, respectively, although the same symbols denote SISO systems in Fig. 4.11. W_s, W_u are also the weighting function matrices associated with the sensitivity function S and the quasi-complementary sensitivity function T_a respectively. In Fig. 4.13, at the top, System configuration 1 contains neither model uncertainty nor W_n; the second one 1b does not contain W_n but the model uncertainty ΔI; in Fig. 4.14 System configuration 2 does not contain model uncertainty but W_n; and the last one 2b contains both the model uncertainty and W_n.

W_s for S is defined depending on the base oscillation models as follows:

$$W_s = \begin{bmatrix} w_s & 0 \\ 0 & w_s \end{bmatrix}, \tag{4.63}$$

$$w_s = \frac{0.4\pi}{s + 10^{-5}} \prod_{i=1}^{n_m} \frac{s^2 + 2\omega_i s + \omega_i^2}{s^2 + 0.02\omega_i s + \omega_i^2}, \tag{4.64}$$

Fig. 4.13 System configurations for $\mathcal{H}_\infty$ control design and analysis without W_n

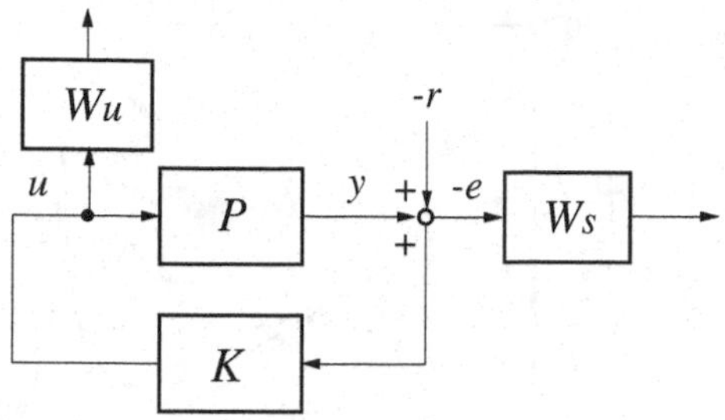

System configuration 1

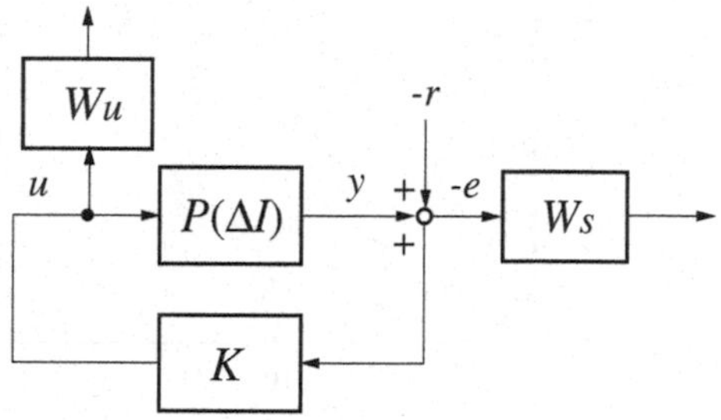

System configuration 1b

Fig. 4.14 System configurations for $\mathcal{H}_\infty$ control design and analysis with W_n

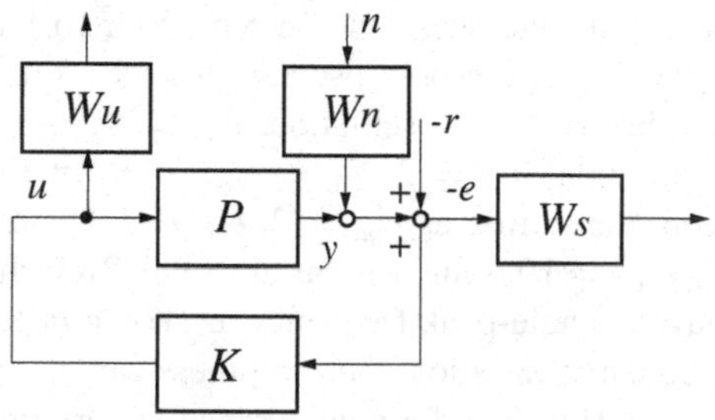

System configuration 2

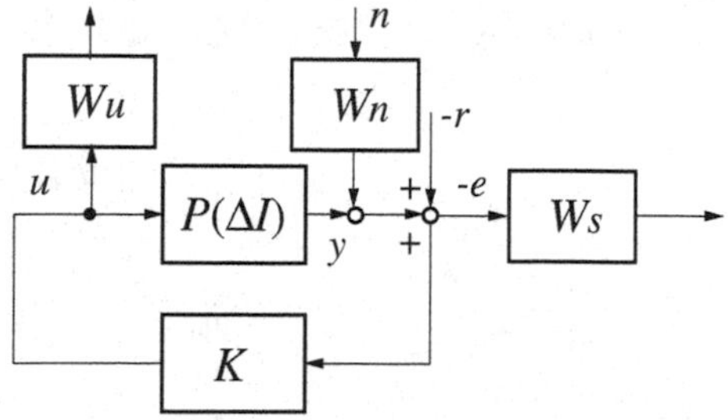

System configuration 2b

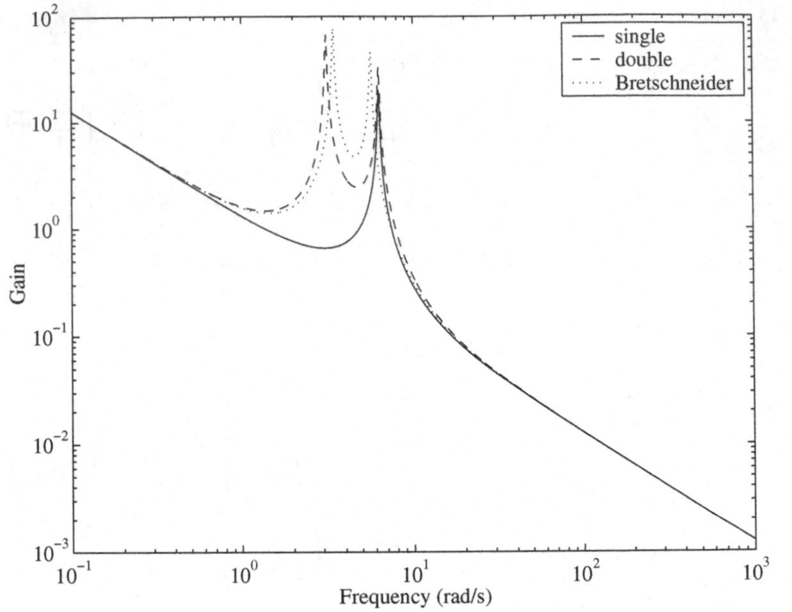

Fig. 4.15 Frequency responses of w_s's

where w_s includes a notch-type filter to let S_p and S have anti ω_i modes corresponding to the base oscillation, and a quasi-integrator for step tracking. For the single-frequency oscillation, $n_m = 1$, $\omega_1 = 2\pi$ rad/s; for the double-frequency oscillation, $n_m = 2$, $\omega_1 = \pi$, $\omega_2 = 2\pi$ rad/s; and for the Bretschneider oscillation, the setting as $n_m = 2$, $\omega_1 = 1.1\pi$, $\omega_2 = 1.7\pi$ rad/s works well regardless of $n_\omega = 10$, which is because the Bretschneider frequencies are very close and have a single-peak frequency as shown in Fig. 2.6. For a small range of oscillation frequency variation, the robustness can be ensured by using multiple ω_i weights as above. However, for a large range of variation, e.g., a case of a vessel with a variety of traveling velocities, some adaptive schemes would be required, which is one of the problems to be tackled in the future. The respective frequency responses of w_s's are shown in Fig. 4.15. Then, W_u and W_n are set as follows:

$$W_u = \begin{bmatrix} w_u & 0 \\ 0 & w_u \end{bmatrix}, \tag{4.65}$$

$$w_u = \frac{10^{-4}s + 10^{-2}}{s + 10^3}, \tag{4.66}$$

$$W_n = \begin{bmatrix} w_n & 0 \\ 0 & w_n \end{bmatrix}, \tag{4.67}$$

$$w_n = \frac{20(s + 10\pi)}{s + 37\pi}. \tag{4.68}$$

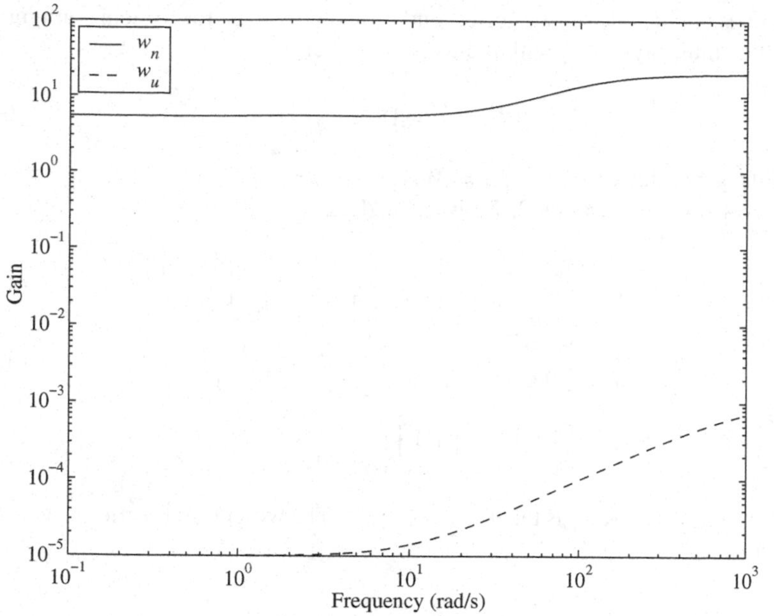

Fig. 4.16 Frequency responses of w_u and w_n

where w_u is a high-pass filter for the control input u and w_n is a high-pass filter for sensor noises and additive model uncertainties. As shown in Fig. 4.16, w_u is nonrestrictive compared with w_n.

Based on the respective system configurations, corresponding generalized plants are constructed. The generalized plants are utilized in $\mathcal{H}_\infty$ controller synthesis and robustness analysis. Controller synthesis methods adopted depends on the system configurations; in the case of System configurations 1 and 2 in Figs. 4.13 and 4.14 which do not contain the model uncertainty ΔI, since the plant is decoupled, "γ-iteration" for a SISO $\mathcal{H}_\infty$ controller is conducted; whereas in the case of the configurations 1b and 2b with ΔI, "D-K iteration" for an MIMO $\mathcal{H}_\infty$ controller is performed.

First, we consider the case of System configurations 1 and 2, where every block consists of decoupled identical SISO systems. Then, we can deal with the system as a SISO system within these frameworks. Hence, the state-space representation of P is given by (4.62). Augmenting P by additional inputs and outputs, the reference signal r_i, the control error e_i of Link i, for performance evaluation, yields

$$\dot{\boldsymbol{x}} = \begin{bmatrix} 0 & 1 \\ -F_1 & -F_2 \end{bmatrix} \boldsymbol{x} + \begin{bmatrix} 0 \\ 1 \end{bmatrix} (-r_i) + \begin{bmatrix} 0 \\ 1 \end{bmatrix} u_i, \tag{4.69a}$$

$$\begin{bmatrix} -e_i \\ u_i \end{bmatrix} = \begin{bmatrix} 1 & 0 \\ 0 & 0 \end{bmatrix} \boldsymbol{x} + \begin{bmatrix} 1 \\ 0 \end{bmatrix} (-r_i) + \begin{bmatrix} 0 \\ 1 \end{bmatrix} u_i, \tag{4.69b}$$

$$y_i = \begin{bmatrix} 1 & 0 \end{bmatrix} \boldsymbol{x} - r_i, \tag{4.69c}$$

which does not include sensor noises, and is denoted by Σ_1 for System configuration 1. Further, multiplying Σ_1 with the weighting matrix

$$W_1 = \mathrm{diag}[w_s, w_u, 1] \tag{4.70}$$

yields the generalized plant $P_{aug1} = W_1 \Sigma_1$.

For System configuration 2, Σ_2 is defined as:

$$\dot{\boldsymbol{x}} = \begin{bmatrix} 0 & 1 \\ -F_1 & -F_2 \end{bmatrix} \boldsymbol{x} + \begin{bmatrix} 0 & 0 \\ 1 & 0 \end{bmatrix} \begin{bmatrix} -r_i \\ n_i \end{bmatrix} + \begin{bmatrix} 0 \\ 1 \end{bmatrix} u_i, \tag{4.71a}$$

$$\begin{bmatrix} -e \\ u_i \end{bmatrix} = \begin{bmatrix} 1 & 0 \\ 0 & 0 \end{bmatrix} \boldsymbol{x} + \begin{bmatrix} 1 & 1 \\ 0 & 0 \end{bmatrix} \begin{bmatrix} -r_i \\ n_i \end{bmatrix} + \begin{bmatrix} 0 \\ 1 \end{bmatrix} u_i, \tag{4.71b}$$

$$y_i = \begin{bmatrix} 1 & 0 \end{bmatrix} \boldsymbol{x} + \begin{bmatrix} 1 & 1 \end{bmatrix} \begin{bmatrix} -r_i \\ n_i \end{bmatrix}, \tag{4.71c}$$

where the sensor noise n_i is taken into account. The weighting function matrices are defined as

$$W_{2l} = \mathrm{diag}[w_s, w_u, 1], \tag{4.72}$$

$$W_{2r} = \mathrm{diag}[1, w_n, 1], \tag{4.73}$$

with which the generalized plant is constructed as $P_{aug2} = W_{2l} \Sigma_2 W_{2r}$.

P_{aug1} and P_{aug2} are employed to synthesize corresponding $\mathcal{H}_\infty$ controllers via γ-iteration.

Next, we address the case of System configurations 1b and 2b with an explicit model uncertainty representation. Figure 4.17 depicts the generalized plant with the dashed line box for System configuration 2b in Fig. 4.14, and the controller to be synthesized K, the normalized structured model uncertainty matrix Δ_I in (4.55), and additionally the fictitious full-block complex uncertainty matrix Δ_p with $\bar{\sigma}(\Delta_p) \leq 1$ intended to accommodate robust performance.

To obtain an LFT representation of $P(\Delta I)$ for System configurations 1b and 2b as in Fig. 4.17, applying the formula of the inverse of a LFT, e.g., [115], to the LFT of $I + \Delta I$ in (4.57) yields

$$\begin{aligned} \mathcal{F}_u(\Phi', \Delta_I) &= (I + \Delta I)^{-1} \\ &= \mathcal{F}_u(\Phi, \Delta_I)^{-1}, \end{aligned} \tag{4.74a}$$

$$\begin{aligned} \Phi' &= \begin{bmatrix} \Phi'_{11} & \Phi'_{12} \\ \Phi'_{21} & \Phi'_{22} \end{bmatrix} \\ &= \begin{bmatrix} \Phi_{11} - \Phi_{12}\Phi_{22}^{-1}\Phi_{21} & -\Phi_{12}\Phi_{22}^{-1} \\ \Phi_{22}^{-1}\Phi_{21} & \Phi_{22}^{-1} \end{bmatrix} \\ &= \left[\begin{array}{cc|c} \mathbf{0} & -(\Delta I_1 - \Delta I_2) & -(\Delta I_1 - \Delta I_2) \\ \frac{1}{2} I_2 & -\frac{1}{2}(\Delta I_1 + \Delta I_2) & -\frac{1}{2}(\Delta I_1 + \Delta I_2) \\ \hline \mathbf{0} & I_2 & I_2 \end{array}\right]. \end{aligned} \tag{4.74b}$$

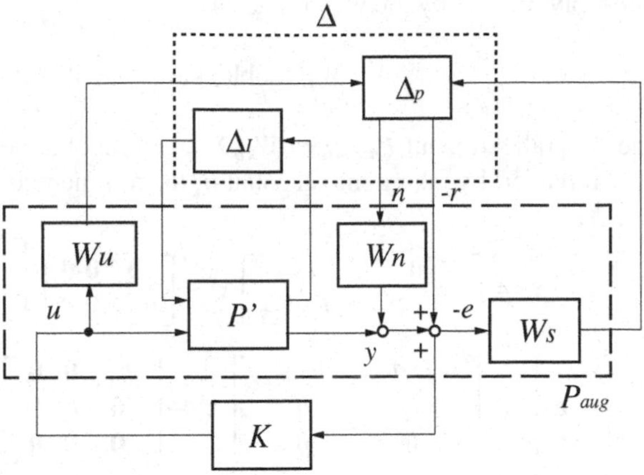

Fig. 4.17 Generalized plant for $\mathcal{H}_\infty$ controller synthesis and robustness analysis based on System configuration 2b

Thus, $P(\Delta I) = \mathcal{F}_u(P', \Delta_I)$ such that the state-space representation of P' is expressed by

$$\dot{x} = \begin{bmatrix} \mathbf{0} & I_2 \\ -F_1\Phi'_{22} & -F_2\Phi'_{22} \end{bmatrix} x + \begin{bmatrix} \mathbf{0} \\ \Phi'_{21} \end{bmatrix} w + \begin{bmatrix} \mathbf{0} \\ \Phi'_{22} \end{bmatrix} u, \tag{4.75a}$$

$$z = \begin{bmatrix} -F_1\Phi'_{12} & -F_2\Phi'_{12} \end{bmatrix} x + \Phi'_{11} w + \Phi'_{12} u, \tag{4.75b}$$

$$y = \begin{bmatrix} I_2 & \mathbf{0} \end{bmatrix} x, \tag{4.75c}$$

where $x = [q_1, q_2, \dot{q}_1, \dot{q}_2]^T$, $y = [q_1, q_2]^T$, $u = [u_1, u_2]^T$, the new control input defined in (4.32), $z \in \mathbb{R}^4$, $w \in \mathbb{R}^4$. With a slight abuse of notation, z and w are related as $w = \Delta_I z$.

Further, by incorporating additional inputs and outputs similarly to P_{aug1} and P_{aug2}, i.e., the reference vector $-r$, the sensor noise vector n, and the control error vector e, the augmented plant for System configuration 1b denoted by Σ_{1b} is expressed as follows:

$$\dot{x} = \begin{bmatrix} \mathbf{0} & I_2 \\ -F_1\Phi'_{22} & -F_2\Phi'_{22} \end{bmatrix} x + \begin{bmatrix} \mathbf{0} & \mathbf{0} \\ \Phi'_{21} & \mathbf{0} \end{bmatrix} \begin{bmatrix} w \\ -r \end{bmatrix} + \begin{bmatrix} \mathbf{0} \\ \Phi'_{22} \end{bmatrix} u, \tag{4.76a}$$

$$\begin{bmatrix} z \\ -e \\ u \end{bmatrix} = \begin{bmatrix} -F_1\Phi'_{12} & -F_2\Phi'_{12} \\ I_2 & \mathbf{0} \\ \mathbf{0} & \mathbf{0} \end{bmatrix} x + \begin{bmatrix} \Phi'_{11} & \mathbf{0} \\ \mathbf{0} & I_2 \\ \mathbf{0} & \mathbf{0} \end{bmatrix} \begin{bmatrix} w \\ -r \end{bmatrix} + \begin{bmatrix} \Phi'_{12} \\ \mathbf{0} \\ I_2 \end{bmatrix} u, \tag{4.76b}$$

$$y = \begin{bmatrix} I_2 & \mathbf{0} \end{bmatrix} x + \begin{bmatrix} \mathbf{0} & I_2 \end{bmatrix} \begin{bmatrix} w \\ -r \end{bmatrix}. \tag{4.76c}$$

Multiplying Σ_{1b} by the weighting function matrix

$$W_{1b} = \text{blockdiag}[I_4, W_s, W_u], \tag{4.77}$$

the generalized plant $P_{aug1b} = W_{1b}\Sigma_{1b}$ can be obtained.

Moreover, by taking into account $\boldsymbol{n}$, the augmented plant Σ_{2b} is given by

$$\dot{\boldsymbol{x}} = \begin{bmatrix} \mathbf{0} & I_2 \\ -F_1\Phi'_{22} & -F_2\Phi'_{22} \end{bmatrix} \boldsymbol{x} + \begin{bmatrix} \mathbf{0} & \mathbf{0}\,\mathbf{0} \\ \Phi'_{21} & \mathbf{0}\,\mathbf{0} \end{bmatrix} \begin{bmatrix} \boldsymbol{w} \\ -\boldsymbol{r} \\ \boldsymbol{n} \end{bmatrix} + \begin{bmatrix} \mathbf{0} \\ \Phi'_{22} \end{bmatrix} \boldsymbol{u}, \tag{4.78a}$$

$$\begin{bmatrix} \boldsymbol{z} \\ -\boldsymbol{e} \\ \boldsymbol{u} \end{bmatrix} = \begin{bmatrix} -F_1\Phi'_{12} & -F_2\Phi'_{12} \\ I_2 & \mathbf{0} \\ \mathbf{0} & \mathbf{0} \end{bmatrix} \boldsymbol{x} + \begin{bmatrix} \Phi'_{11} & \mathbf{0} & \mathbf{0} \\ \mathbf{0} & I_2 & I_2 \\ \mathbf{0} & \mathbf{0} & \mathbf{0} \end{bmatrix} \begin{bmatrix} \boldsymbol{w} \\ -\boldsymbol{r} \\ \boldsymbol{n} \end{bmatrix} + \begin{bmatrix} \Phi'_{12} \\ \mathbf{0} \\ I_2 \end{bmatrix} \boldsymbol{u}, \tag{4.78b}$$

$$\boldsymbol{y} = \begin{bmatrix} I_2 & \mathbf{0} \end{bmatrix} \boldsymbol{x} + \begin{bmatrix} \mathbf{0} & I_2 & I_2 \end{bmatrix} \begin{bmatrix} \boldsymbol{w} \\ -\boldsymbol{r} \\ \boldsymbol{n} \end{bmatrix}. \tag{4.78c}$$

Then, defining weighting function matrices as

$$W_{2bl} = \text{blockdiag}[I_4, W_s, W_u], \tag{4.79}$$

$$W_{2br} = \text{blockdiag}[I_6, W_n], \tag{4.80}$$

the generalized plant for System configuration 2b is acquired as $P_{aug2b} = W_{2bl}\Sigma_{2b}W_{2br}$.

Consequently, being based on the obtained generalized plants P_{aug1b} or P_{aug2b}, and employing the corresponding structured uncertainty matrix

$$\Delta = \text{blockdiag}[\Delta_I, \Delta_p], \tag{4.81}$$

D–K iteration is performed to synthesize an $\mathcal{H}_\infty$ controller. Depending on the generalized plants, the dimension of Δ is defined as in the following:

$$\begin{aligned} \boldsymbol{\Delta}_1 := \{&\text{blockdiag}[\delta_1(t)I_2, \delta_2 I_2, \Delta_P] : \\ &\delta_i \in \mathbb{R}, |\delta_i| \leq 1, \Delta_P \in \mathbf{C}^{2\times 4}, \bar{\sigma}(\Delta_P) \leq 1\}. \end{aligned} \tag{4.82}$$

for P_{aug1b}, and

$$\begin{aligned} \boldsymbol{\Delta}_2 := \{&\text{blockdiag}[\delta_1(t)I_2, \delta_2 I_2, \Delta_P] : \\ &\delta_i \in \mathbb{R}, |\delta_i| \leq 1, \Delta_P \in \mathbf{C}^{4\times 4}, \bar{\sigma}(\Delta_P) \leq 1\}. \end{aligned} \tag{4.83}$$

for P_{aug2b} respectively, where $\boldsymbol{\Delta}_1$ and $\boldsymbol{\Delta}_2$ are the uncertainty sets which possible Δ belongs to.

4.4.5 $\mathcal{H}_\infty$ Controller Synthesis and Robustness Analysis

4.4.5.1 $\mathcal{H}_\infty$ Controller Synthesis

Now we demonstrate $\mathcal{H}_\infty$ controller synthesis and robustness analysis using the constructed generalized plants. As mentioned above, being based on P_{aug1} or P_{aug2} without explicit uncertainty model, γ-iteration is conducted to obtain an SISO $\mathcal{H}_\infty$ controller, whereas for P_{aug1b} or P_{aug2b} with consideration of the explicit uncertainty model, D–K iteration is done getting an MIMO $\mathcal{H}_\infty$ one. For each generalized plant P_{augi} $(i = 1, 1b, 2, 2b)$, three controllers denoted by K_{i1}, K_{i2}, K_{i3} are synthesized according to the respective oscillation models, the single-frequency one, the double-frequency one, the Bretschneider one.

Then, γ-iteration with P_{augi} $(i = 1, 2)$ has given the resultant $\mathcal{H}_\infty$ controller denoted by K_{ij} $(j = 1, 2, 3)$ with the upper bound of γ as

$$||\mathcal{F}_l(P_{augi}, K_{ij})||_\infty \le \gamma_{ij}, \tag{4.84}$$

where γ_{ij} is shown in Table 4.2. These values of γ_{ij}'s being less than one imply that all the controllers can provide better performance than the respective specified ones in the nominal model case, and the differences between γ_{1j}'s and γ_{2j}'s show that the weighting function w_n is restrictive to an certain degree. Note that this γ-iteration-based synthesis is equivalent to a synthesis to obtain a robust stabilizing controller with an unstructured complex uncertainty.

Next, we present the results of D–K iteration using P_{augi} $(i = 1b, 2b)$ and the corresponding structured uncertainty matrices $\Delta \in \mathbf{\Delta}_1$ in (4.82) and $\Delta \in \mathbf{\Delta}_2$ in (4.83), and this synthesis procedure means to solve the robust performance problem. The resultant upper μ bound for P_{augi} with the index j representing the oscillation model as above has been obtained as

$$\sup_{\omega \in [0,\infty]} ||D_l(i\omega)\mathcal{F}_l(P_{augi}(i\omega), K_{ij}(i\omega))D_r(i\omega)||_\infty \le \mu_{ij}, \tag{4.85}$$

where ω is the frequency variable and D_l and D_r are the frequency-dependent complex scaling matrices which satisfy that $D_r\Delta = \Delta D_l^{-1}$. The dimensions and structures of D_l and D_r depend on P_{augi}; specifically, the set of D_l and D_r involved for P_{aug1b} is as follows:

Table 4.2 Achievable upper γ bound in each SISO $\mathcal{H}_\infty$ controller synthesis via γ-iteration; "single," "double," "Bretschneider" represent the single-frequency oscillation, the double-frequency one, and the Bretschneider one, respectively

P_{aug}	Single	Double	Bretschneider
P_{aug1}	0.156	0.156	0.156
P_{aug2}	0.703	0.859	0.859

Table 4.3 Achievable upper μ bound in each MIMO $\mathcal{H}_\infty$ controller synthesis via D–K iteration; "single", "double", "Bretschneider" represent the single-frequency oscillation, the double-frequency one, and the Bretschneider one, respectively

P_{aug}	Single	Double	Bretschneider
P_{aug1b}	0.633	0.676	0.873
P_{aug2b}	1.601	1.479	2.019

$$\mathbf{D} = \left\{ \begin{array}{l} (D_l, D_r) : D_l = \text{blockdiag}[D_1, D_2, d_1 I_4], \\ D_r = \text{blockdiag}[D_1^{-1}, D_2^{-1}, d_1^{-1} I_2], \\ D_i = D_i^* \succ 0 \in \mathbf{C}^{2\times 2}, d_1 > 0 \in \mathbb{R} \end{array} \right\}, \tag{4.86}$$

and the set for P_{aug2b} is as follows:

$$\mathbf{D} = \left\{ \begin{array}{l} (D_l, D_r) : D_l = \text{blockdiag}[D_1, D_2, d_1 I_4], \\ D_r = \text{blockdiag}[D_1^{-1}, D_2^{-1}, d_1^{-1} I_4], \\ D_i = D_i^* \succ 0 \in \mathbf{C}^{2\times 2}, d_1 > 0 \in \mathbb{R} \end{array} \right\}. \tag{4.87}$$

However, in practice, μ_{ij} is computed over finite points logarithmically spaced in a designated frequency range $I_\omega = \{\omega_1, \omega_2, \ldots, \omega_n\}$ as

$$\mu_{ij} = \max_{\omega_k \in I_\omega} ||D_l(i\omega_k)\mathcal{F}_l(P_{augi}(i\omega_k), K_{ij}(i\omega_k))D_r(i\omega_k)||_\infty, \tag{4.88}$$

where in our demonstration 100 points from 10^{-1} to 10^3 rad/s have been employed as I_ω. Table 4.3 shows the obtained μ_{ij}'s. μ_{1bj}'s ($j = 1, 2, 3$) are less than one, which implies that the controllers K_{1bj}'s can achieve the desired robust performance without considering W_n if Δ is time-invariant. However, since Δ is time-varying uncertainty, strictly speaking, these upper μ bounds cannot give us definite evaluation in terms of robust performance or robust stability either, but some reasonable properties can be expected. Then, since P_{aug2b} contains W_n, μ_{2bj}'s are greater than one, but not that large. Similarly, the results do not present any conclusive facts either, however, feasible controllers can be expected. At least with respect to robust stability, more strict analysis is necessary, which will be demonstrated later.

Here it should be noted that a controller synthesized via D–K iteration, in general, results in a considerably high-order system due to dynamical system scalings. In our demonstration case, the order of each K_{ij} is shown in Table 4.4, where K_{1j}'s and K_{2j}'s are SISO controllers, while K_{1bj}'s and K_{2bj} are MIMO ones. Since the SISO controllers are employed as follows:

$$\text{diag}[K_{ij}, K_{ij}], \tag{4.89}$$

whose order amounts to the double of the order of K_{ij}. As seen from the table, the orders of K_{1bj} and K_{2bj} are too large compared to ones of the controllers obtained

Table 4.4 Orders of the obtained $\mathcal{H}_\infty$ controllers

Controller	Order	Controller	Order	Controller	Order
K_{11}	6	K_{12}	8	K_{13}	8
K_{1b1}	92	K_{1b2}	96	K_{1b3}	96
K_{1b1red}	12	K_{1b2red}	16	K_{1b3red}	16
K_{21}	7	K_{22}	9	K_{23}	9
K_{2b1}	94	K_{2b2}	98	K_{2b3}	98
K_{2b1red}	14	K_{2b2red}	18	K_{2b3red}	18

Table 4.5 Zeros and poles of the nominal plant and weighting functions

Transfer function	Zeros	Poles
P		−2.083e+1
		−3.168e+0
W_s	−6.283e+0	−6.283e+0 + 6.283e+0i
	−6.283e+0	−6.283e+0 − 6.283e+0i
		−1.000e-5
W_u	−1.000e+1	−1.000e+3
W_n	−3.142e+1	−1.162e+2

via γ-iteration. Hence, we have performed model reduction to those controllers by using the balanced realization with Hankel singular values. Each reduced order was determined so as to be consistent with that of the corresponding controller obtained via γ-iteration. The resultant controllers through model reduction are denoted by adding the suffix "*red*," e.g., from K_{1b1} to K_{1b1red}. For ease of implementation and analysis, the reduced-order controllers will be mainly used instead of the original ones.

Then let us investigate the obtained controllers in some detail, particularly by considering the case of single-frequency oscillation for exposition simplicity. First, we focus on zeros and poles of the obtained controllers K_{i1}'s. Table 4.5 shows zeros and poles of the nominal plant and weighting functions, which are reflected to those of the $\mathcal{H}_\infty$ controllers as shown in Table 4.6. K_{11} and K_{21} contain all the poles of P, W_u as zeros, and K_{21} has additionally the pole of W_n as a zero. While $K_{1bred22}$ and $K_{2bred22}$ contain similar zeros, which however have slightly shifted from the poles by taking into account of the model uncertainties. Considering poles of the controllers, we see that all the controllers contain all the poles of W_s, i.e., the pole for quasi-integrator and ones of the base oscillation frequency ω_1, as poles which have not shifted at all even under consideration of model uncertainties. This fact represents a considerably important feature of $\mathcal{H}_\infty$ control which ensures a systematic control design to implement so-called "the internal model principle" by modifying W_s. As the above argument, in general, an $\mathcal{H}_\infty$ controller tries to shape the objective transfer functions by adopting zeros or poles from the poles of the given plant and weighting functions, which can provide a control design perspective [94, 95].

Table 4.6 Zeros and poles of the obtained $\mathcal{H}_\infty$ controllers

Controller	Zeros	Controller	Zeros
K_{11}	−1.000e+3	K_{21}	−1.000e+3
	−2.083e+1		−2.083e+1
	−3.168e+0		−3.168e+0
	−3.103e+1 + 3.450e+0i		−3.025e+0 + 3.421+0i
	−3.103e+1 − 3.450e+0i		−3.025e+0 − 3.421e+0i
			−1.112 e + 2
$K_{1b1red22}$	−9.409e+2	$K_{2bred22}$	−1.022e+3
	−2.656e+1		−2.173e+1
	−3.210e+0		−3.119e+0
	−7.190e-1 + 3.411e+0i		−1.979e+0 + 3.468e+0i
	−7.190e-1 − 3.411e+0i		−1.979e+0 − 3.468e+0i
	−1.964 e+2		−1.413e+2 + 1.273e+0i
			−1.413e+2 − 1.273e+0i
Controller	Poles	Controller	Poles
K_{11}	−1.000e-5	K_{21}	−9.992e-6
	−6.283e-2 + 6.283e+0i		−6.283e+0 + 6.283e+0i
	−6.283e-2 − 6.283e+0i		−6.283e+0 − 6.283e+0i
	−2.597e+1 + 4.891e+1i		−3.253e+0 + 4.719e+1i
	−2.597e+1 − 4.891e+1i		−3.253e+0 − 4.719e+1i
	−6.768e+1		−6.388e+1
			−6.727e+2
$K_{1b1red22}$	−1.000e-5	$K_{2bred22}$	−1.000e-5
	−6.283e-2 + 6.283e+0i		−6.282e-2 + 6.283e+0i
	−6.283e-2 − 6.283e+0i		−6.282e-2 − 6.283e+0i
	−3.071e+1 + 2.674e+1i		−2.867e+1 + 3.993e+1i
	−3.071e+1 − 2.674e+1i		−2.867e+1 − 3.993e+1i
	−4.415e+2		−2.160e+2
	−6.703e+1		−7.729e+1 + 1.554e+1i
			−7.729e+1 − 1.554e+1i

For the reduced-order controllers, the pole-zero cancelation pairs are omitted

Next, we present frequency responses of the controllers and the resultant sensitivity functions of interest in (4.59)–(4.61). Figure 4.18 depicts those of the $\mathcal{H}_\infty$ controllers. All the controllers have the resonant mode of ω_1 and the quasi-integrator property given by W_s, which ensures the internal model property for tracking control and disturbance attenuation for step signals and ones of frequencies near ω_1. Comparing the controllers, it is seen that in the frequency range where tracking control and disturbance attenuation are required, K_{11} is the highest gain controller and K_{12} is almost the same as it. However, in high frequencies of greater than 70 rad/s, the gain of K_{12} is higher than that of K_{11}, which is effect of W_n restricting S in (4.59). Addition of model uncertainties has resulted in relatively low-gain controllers so as to attain robustness as expected, particularly K_{1b1red} is the lowest gain one, but addition of W_n has increased the gain of K_{2b1red} also by restricting S. Consider difference between the channels in K_{1b1red} and K_{2bred} in Fig. 4.18. The diagonal element $K_{1b1redii}$ and $K_{2b1redii}$ $(i = 1, 2)$ are the controllers for Joint i. Comparing the controllers for Joints 1 and 2, the gains of the controllers for Joint 1 are lower than the corresponding ones for Joint 2, which reflects the fact that control of Joint 1 requires more robustness due to inertia parameter variation due to change of q_2.

Figure 4.19 shows the frequency responses of the resultant sensitivity functions with the respective $\mathcal{H}_\infty$ controllers. In addition to S, S_p, and T_a in (4.59)–(4.61), the complimentary sensitivity function $T := PT_a$ (for multiplicative model uncertainties) is depicted. The sensitivity functions with K_{11} (solid line) and K_{21} (dashed line) reveal desirable profiles as depicted in Fig. 4.12. Due to the difference in the gain profiles of those controllers, the gains of S and S_p with K_{21} are slightly lower than those with K_{11} in frequencies of less than 70 rad/s, while it is the opposite case in frequencies of greater than 70 rad/s, which is difficult to be seen on a normal scale, but it will be possible by magnifying the graph. This difference implies that, in the nominal case, K_{11} will perform slightly better tracking control and disturbance attenuation than K_{21} while K_{21} will exhibit less influence of high-frequency sensor noise than K_{11}. On the other hand, T and T_a with K_{11} and K_{21} directly reflect the controller gains, particularly in frequencies of greater than 70 rad/s.

Then let us see the cases of K_{1b1red} and K_{2b1red}. Similarly with the above cases, the gain profiles of the respective sensitivity functions well reflect the properties of the controllers. That is, since the gains of S and S_p with K_{1b1red} are higher than those with the other controllers due to consideration of model uncertainties, in the nominal case, K_{1b1red} will exhibit poorer performance of tracking control and disturbance attenuation. Further, the profiles with K_{11} reveals unfavorable deformed shape in frequencies near ω_1. The property of K_{2b1red} has been slightly recovered by W_n. However, these two controllers K_{1b1red} and K_{2b1red} can be expected to perform better than K_{11} and K_{21} in the worst case of model variation, which will be demonstrated later. Moreover, we present the resultant sensitivity functions in the cases of double-frequency oscillation and Bretschneider oscillation in Figs. 4.20 and 4.21 respectively. It should be noted that the essential properties and arguments described above are common ones for all the frequency cases.

4.4.5.2 Robustness Analyses

Here we perform robustness analyses on the resultant control systems in terms of stability and performance. We will employ two types of approaches, one of which is based on μ analysis, i.e., the scaled $\mathcal{H}_\infty$-norm approach, and the other one utilizes the state-dependent coefficient (SDC) form with the Lyapunov theory.

First, we present μ-based robustness analysis. As mentioned in the previous $\mathcal{H}_\infty$ controller synthesis section, conventional μupper bounds with frequency-dependent scalings $D(i\omega)$'s can accommodate only time-invariant uncertainties properly. For time-varying uncertainties as in our case, constant scalings over all the frequencies

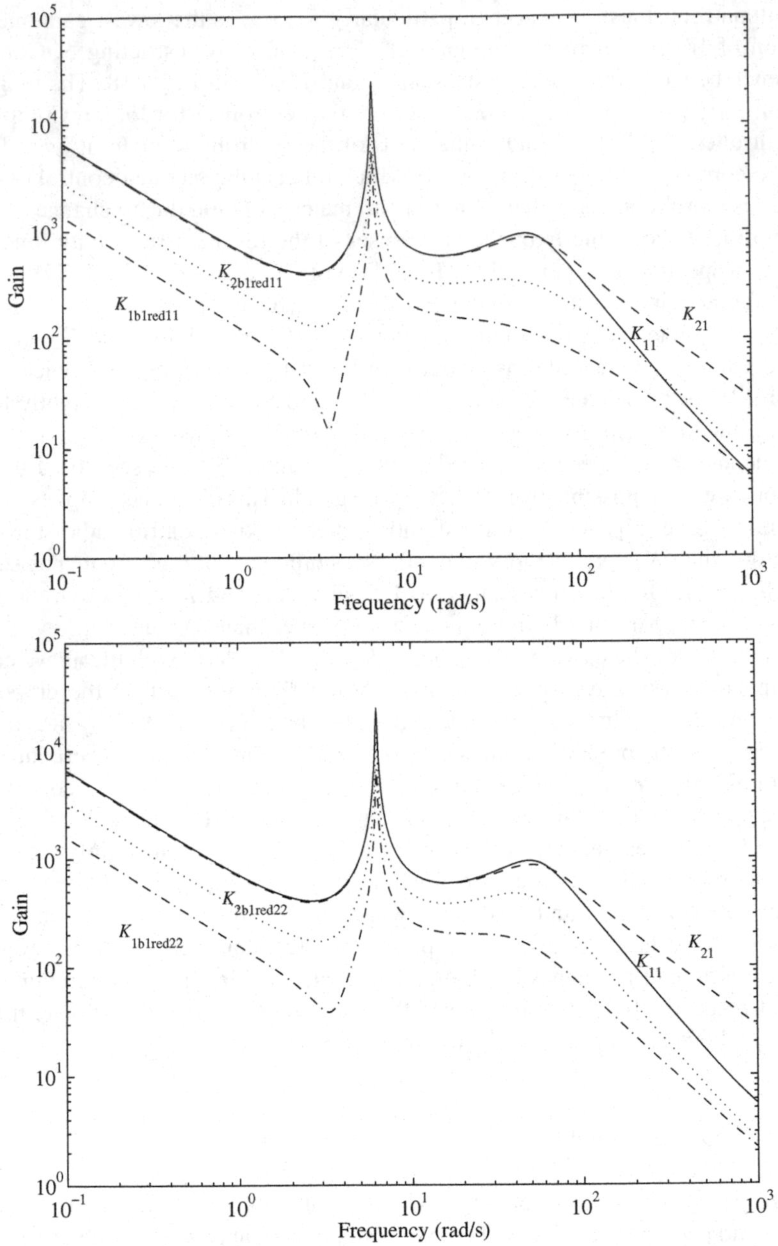

Fig. 4.18 Frequency responses of the $\mathcal{H}_\infty$ controllers; *solid line* K_{11}, *dashed line* K_{21}, *dashed-dotted line* $K_{1b1red11}$ (*top*) and $K_{1b1red22}$ (*bottom*), *dotted line* $K_{2b1red11}$ (*top*) and $K_{2b1red22}$ (*bottom*)

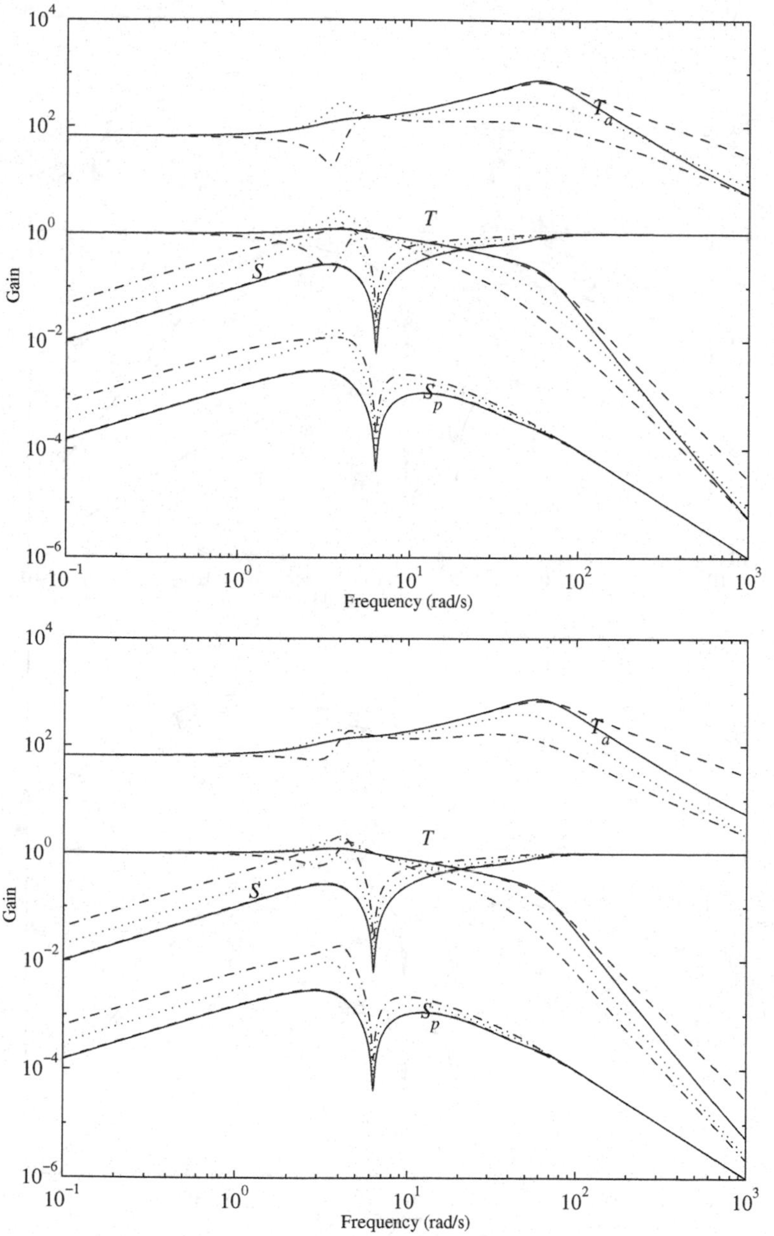

Fig. 4.19 Frequency responses of the resultant sensitivity functions with the respective $\mathcal{H}_\infty$ controllers (single-frequency case); *solid line* K_{11}, *dashed line* K_{21}, *dashed-dotted line* $K_{1b1red11}$ (*top*) and $K_{1b1red22}$ (*bottom*), *dotted line* $K_{2b1red11}$ (*top*) and $K_{2b1red22}$ (*bottom*)

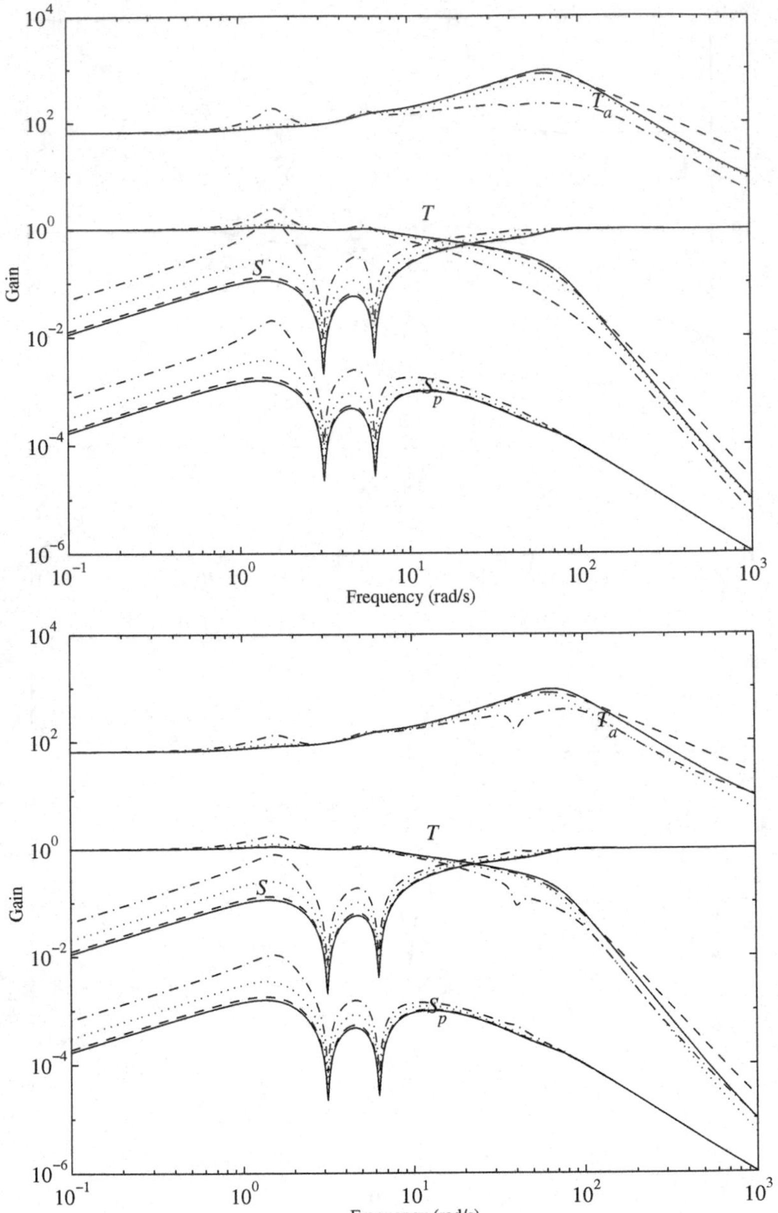

Fig. 4.20 Frequency responses of the resultant sensitivity functions with the respective $\mathcal{H}_\infty$ controllers (double-frequency case); *solid line* K_{11}, *dashed line* K_{21}, *dashed-dotted line* $K_{1b1red11}$ (*top*) and $K_{1b1red22}$ (*bottom*), *dotted line* $K_{2b1red11}$ (*top*) and $K_{2b1red22}$ (*bottom*)

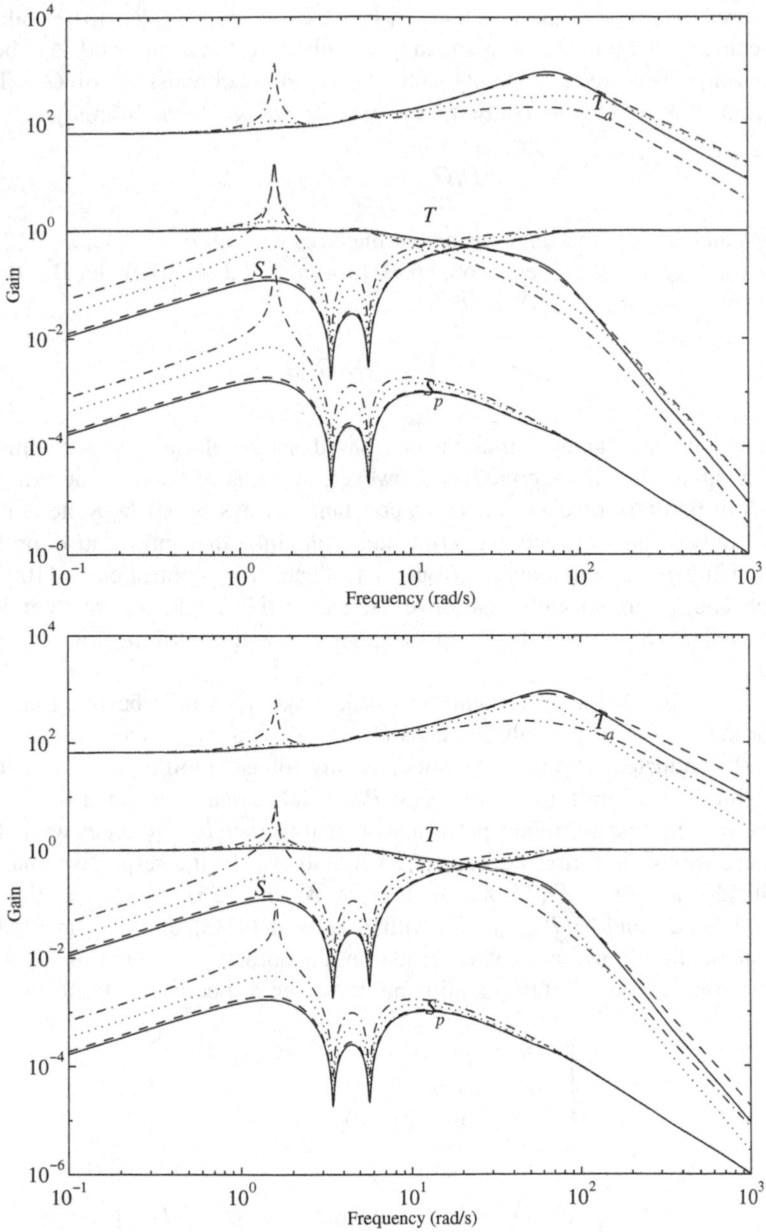

Fig. 4.21 Frequency responses of the resultant sensitivity functions with the respective $\mathcal{H}_\infty$ controllers (Bretschneidr case); *solid line* K_{11}, *dashed line* K_{21}, *dashed-dotted line* $K_{1b1red11}$ (*top*) and $K_{1b1red22}$ (*bottom*), *dotted line* $K_{2b1red11}$ (*top*) and $K_{2b1red22}$ (*bottom*)

must be used for stringent analyses. Let G_{ev} be the closed-loop system to be evaluated which contains the plant, controller, maybe weighting functions, and Δ_{ev} be the time-varying normalized uncertainty with the compatible dimension with G_{ev}. Then, $\mathcal{F}_u(G_{ev}, \Delta_{ev})$ is stable if and only if there exist $D_l, D_r \in \mathbb{R}$ which satisfy

$$\|D_l G_{ev} D_r\|_\infty < 1 \tag{4.90}$$

where D_l and D_r are constant nonsingular matrices with $D_r \Delta_{ev} = \Delta_{ev} D_l^{-1}$ ($\|\cdot\|_\infty$ denotes the $\mathcal{H}_\infty$-norm)(see, e.g., [68, 86, 93] for more detail). Then, let $\mathbf{D}_{ad}$ denote the admissible set of (D_l, D_r) and define

$$\mu_c := \inf_{(D_l, D_r) \in \mathbf{D}_{ad}} \|D_l G_{ev} D_r\|_\infty. \tag{4.91}$$

Therefore, robustness analysis must be done based on the criteria μ_c to accommodate time-varying structured uncertainties. However, in terms of "control design," it is generally difficult to solve a problem via constant scalings-based D–K iteration [3, 79, 109] rather than conventional D–K iteration since the optimization problem of minimizing μ_c is non-convex for output feedback $\mathcal{H}_\infty$ control cases [109] and constant scalings are strongly restrictive. Hence, in this work, we have employed conventional D–K iteration for control design, but will deploy μ_c for robustness analyses.

Regarding the obtained 12 controllers K_{ij}, three types of robustness analyses based on the respective generalized plants P', P_{aug1b}, and P_{aug2b} (see the Sect. 4.4.4 again) are conducted. P' can accommodate only robust stability analysis without any weighting functions for performances; P_{aug1b} additionally contains W_s and W_u, and hence accommodates robust performance; and further P_{aug2b} copes with all the points, i.e., robust performance with W_s, W_u, and W_n. In the respective analyses, G_{ev} with Δ_{ev} are set as $\mathcal{F}_l(P', K_{ij})$ with $\Delta_I \in \mathbf{\Delta}_I$ in (4.55), $\mathcal{F}_l(P_{aug1b}, K_{ij})$ with $\Delta \in \mathbf{\Delta}_1$ in (4.82), and $\mathcal{F}_l(P_{aug2b}, K_{ij})$ with $\Delta \in \mathbf{\Delta}_2$ in (4.83), and μ_c's are obtained by a numerical optimization method based on a quadratic programming algorithm with the initial D_l and D_r of the identity matrices. The respective admissible sets are

$$\mathbf{D}' := \left\{ \begin{array}{l} (D_l, D_r) : D_l = \text{blockdiag}[D_1, D_2], \\ D_r = \text{blockdiag}[D_1^{-1}, D_2^{-1}], \\ D_i (\det D_i \neq 0) \in \mathbb{R}^{2\times 2} \end{array} \right\} \tag{4.92}$$

for P',

$$\mathbf{D}_1 := \left\{ \begin{array}{l} (D_l, D_r) : D_l = \text{blockdiag}[D_1, D_2, d_1 I_4], \\ D_r = \text{blockdiag}[D_1^{-1}, D_2^{-1}, d_1^{-1} I_2], \\ D_i (\det D_i \neq 0) \in \mathbb{R}^{2\times 2}, d_1 (\neq 0) \in \mathbb{R} \end{array} \right\} \tag{4.93}$$

Table 4.7 Results of μ_c with P' for robust stability analyses against Δ_I

j	K_{1j}	K_{1bj}	K_{21}	K_{2bj}
1	0.5923	0.3693	0.5169	0.4780
		0.3693[a]		0.4779[a]
2	0.6500	0.4645	0.5788	0.5693
		0.4938[a]		0.5732[a]
3	0.6490	0.4640	0.5820	0.5315
		0.4734[a]		0.5420[a]

1 single-frequency, *2* double-frequency, *3* Bretschneider.
[a] Denotes the one of the reduced-order controller

for P_{aug1b}, and

$$\mathbf{D}_2 := \left\{ \begin{array}{l} (D_l, D_r) : D_l = \text{blockdiag}[D_1, D_2, d_1 I_4], \\ D_r = \text{blockdiag}[D_1^{-1}, D_2^{-1}, d_1^{-1} I_4], \\ D_i (\det D_i \neq 0) \in \mathbb{R}^{2\times 2}, d_1 (\neq 0) \in \mathbb{R} \end{array} \right\} \tag{4.94}$$

for P_{aug2b}.

Table 4.7 shows the results of μ_c's with P' for robust stability analyses against normalized time-varying structured model uncertainty Δ_I in (4.55), where not only the reduced-order controllers labeled with $_{red}$ but also the original controllers are considered. First, it should be emphasized that all the μ's are less than one, which implies that all the controllers can ensure at least robust stability against Δ_I. Then, we compare μ_c's of K_{1j}'s and K_{1bj}'s ($j = 1, 2, 3$) and notice that μ_c's of K_{1bj}'s are less than those of K_{1j}'s. This fact reflects explicit consideration of model uncertainties via extended matrix polytopes, and shows that the proposed model uncertainty representation method works well. Further, the cases of K_{2j}'s and K_{2bj}'s are similar, but the differences of μ_c are smaller than those in the cases of K_{1j}'s and K_{1bj}'s, which may be due to some conflict between the model uncertainties and the weighting function W_n. Next, consider the difference between μ_c's of K_{1j}'s and K_{2j}'s. μ_c's of K_{2j}'s are smaller than those of K_{1j}'s, which is because W_n influences T_a as well as S and T_a is associated with additive model uncertainties. Moreover, let us compare K_{ibj}'s with the corresponding reduced-order controllers K_{ibjred}'s. In the case of single-frequency oscillation, their μ_c's are almost the same, however μ_c's of the reduced-order controllers are slightly larger than those of the original controllers. Which shows that the model reduction process has decreased the robustness of the controllers very slightly, but almost conserved it.

Table 4.8 presents the results of μ_c's with P_{aug1b} without W_n for robust performance analyses against $\Delta \in \mathbf{\Delta}_1$. In the robust performance analyses, since it was hard to obtain μ_c's of K_{ibj}'s, only those of K_{ibjred}'s are displayed. Not as in the case of robust stability analyses, all the μ_c's are greater than one. However, which is not a problem, because we have already confirmed that all the controllers can provide robust stability. Further, those μ_c's are close to one, thus their robust performance can be expected. With respect to μ_c's relationship among the controllers, the same arguments can be applied as the above robust stability analyses.

Table 4.8 Results of μ_c with P_{aug1b} without W_n for robust performance analyses against $\Delta \in \mathbf{\Delta}_1$

j	K_{1j}	K_{1bjred}	K_{21}	K_{2bjred}
1	1.3431	1.0094	1.2373	1.0709
2	1.5621	1.4173	1.4951	1.4333
3	1.5502	1.3088	1.4884	1.3866

1 single-frequency, *2* double-frequency, *3* Bretschneider

Table 4.9 Results of μ_c with P_{aug2b} with W_n for robust performance analyses against $\Delta \in \mathbf{\Delta}_2$

j	K_{1j}	K_{1bjred}	K_{21}	K_{2bjred}
1	3.3209	3.4501	3.1337	2.7103
2	4.0071	4.3391	3.8031	3.9539
3	4.0114	4.3496	3.8881	3.7440

1 single-frequency, *2* double-frequency, *3* Bretschneider

Finally, the results of μ_c's with P_{aug2b} with W_n for robust performance analyses against $\Delta \in \mathbf{\Delta}_2$ are displayed in Table 4.9. Due to the restrictive W_n, all the μ_c's are greater than those in the above cases. The μ_c's relationship among the controllers reveals different features from those in the above cases. One feature is that μ_c's of K_{1j}'s and K_{1bjred}'s are greater than those of K_{2j}'s and K_{2bjred}'s, which is due to lacking of consideration of W_n. The other feature is that μ_c's of K_{1bjred}'s are greater than those of K_{1j}'s, which is not the case for the above cases, and implies some conflict between W_n and the model uncertainties. On the other hand, μ_c's of K_{2bjred}'s are less than those of K_{2j}'s with one exception of the double-frequency cases. Moreover, K_{2bjred}'s can be expected to have good robust performance in such a situation where model uncertainties and sensor noises need to be taken into account.

From all the above arguments, as long as the μ-based robust analysis is concerned, we have concluded as follows:

- The proposed extended matrix-polytope-based model uncertainty representation is feasible for both control design and robustness analyses;
- As a $\mathcal{H}_\infty$ synthesis method with consideration of structured uncertainties, the conventional D–K iteration is a useful and systematic way even if the uncertainties are real and/or time-varying ones;
- Further, in such a case, the combination of the conventional D–K iteration and μ_c-based robustness analyses is an effective approach;
- W_n and model uncertainties might cause some conflict, thus the balance of the gains of W_n and W_s will be important in a control design procedure.

Now we proceed to state-dependent coefficient (SDC) form-based robust stability analyses. The control system is constructed based on the virtual linear system in (4.62), its stability is thus guaranteed for the linear system. However, once the payload varies, the system is no longer linear. In the above μ-based analyses, the system is considered the linear system with time-varying uncertainties. Here, we address another viewpoint where the system is regarded as nonlinear focusing on its semi-global internal stability, and apply a method that is motivated by [13].

As in [13], we utilize a SDC form as $\dot{x} = A(x)x$. When a state-space realization of the mth-order $\mathcal{H}_\infty$ controller is represented by

$$\dot{\xi} = A_c\xi + B_c q \tag{4.95}$$
$$u = C_c\xi, \tag{4.96}$$

the SDC form of the closed-loop control system with the state $x^T = [q, \dot{q}, \xi]^T$ can be expressed as follows:

$$\begin{bmatrix} \dot{q} \\ \ddot{q} \\ \dot{\xi} \end{bmatrix} = \begin{bmatrix} 0 & I_2 & 0 \\ A_{21} & A_{22} & A_{23} \\ B_c & 0 & A_c \end{bmatrix} \begin{bmatrix} q \\ \dot{q} \\ \xi \end{bmatrix} \tag{4.97}$$
$$A_{21} = M(q)^{-1}\left(-F_1 M_n(q) - \Delta G(q)\right) \tag{4.98}$$
$$A_{22} = M(q)^{-1}\left(-F_2 M_n(q) - \Delta C(q, \dot{q})\right) \tag{4.99}$$
$$A_{23} = M(q)^{-1} M_n(q) C_c \tag{4.100}$$

where M_n is the inertia matrix with the nominal parameters, $[F_1, F_2]$ are linear gains in (4.62), $\Delta G(q)q$ and $\Delta C(q, \dot{q})\dot{q}$ denote the gravitational term and Coriolis–centripetal term perturbations respectively.

Using this SDC form, we first investigated the eigenvalues λ_i's of $A(x)$ over a state-space region Ω. Taking into account of simulation and experimental cases for demonstration, we designate Ω as

$$q \times \dot{q} \times \xi = [-\pi/2, \pi/2] \times [-\pi, \pi] \times [-5, 5] \times [-12, 12] \times \mathbb{R}^m. \tag{4.101}$$

Then, we calculated the maximum of the real part of the eigenvalues

$$\max_{x \in \Omega, i} Re(\lambda_i(A(x))) \tag{4.102}$$

for a given payload width, the results of which are shown by the top graph in Fig. 4.22. In Fig. 4.22, each graph of oscillation case compares the respective controllers. As seen from the graph, for every payload width, the maximum $Re(\lambda_i(A(x)))$ is negative, which indicates that the control system is possibly internally stable over Ω for the payload variations. Further, interestingly enough, K_{1j} and K_{2j} via γ-iteration exhibit almost the same profiles which vary depending on the payload widths, whereas $Re(\lambda_i(A(x)))$'s with K_{1bjred} and K_{2bjred} via D–K iteration are greater than those with K_{1j} and K_{2j}, and invariant with the payload width, that is, some robustness. Strictly speaking, this condition is not sufficient for its robust stability. Cloutier et al. [13] presented a sufficient condition for the internal stability only for the special SDC form with a symmetric $A(x)$, which is however not the general case. Hence, here we construct a new machinery that is applicable to more general SDC forms including ours, which is described by the following theorem.

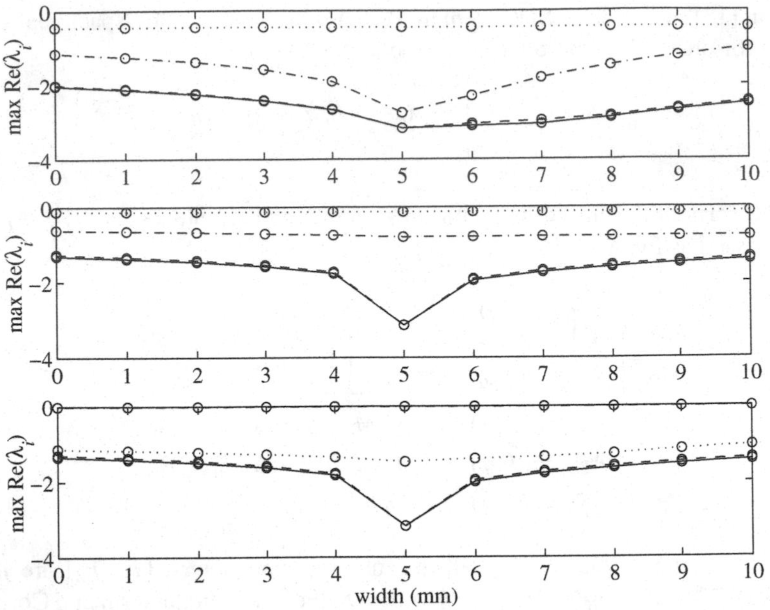

Fig. 4.22 Results of robust stability analyses in terms of maximum $Re(\lambda_i(A(\boldsymbol{x})))$ over Ω in (4.101) versus payload width; *Top* single-frequency oscillation, *Middle* double-frequency oscillation, *Bottom* Bretschneider oscillation; *solid line* K_{1j}, *dashed line* K_{2j}, *dashed-dotted line* K_{1bjred}, *dotted line* K_{2bjred}

Theorem 4.5 *For a given system represented by $\dot{\boldsymbol{x}} = A(\boldsymbol{x})\boldsymbol{x}$, suppose that $A(0)$ is diagonalizable, then there exists a constant nonsingular matrix U that diagonalizes $A(0)$ by $UA(0)U^{-1}$. If U is a complex matrix, then U is replaced by $Re(U)+Im(U)$. For a given state-space region Λ, if*

$$\max_{\boldsymbol{x}\in\Lambda,i} \lambda_i\left((U^{-1})^T A(\boldsymbol{x})^T U^T + UA(\boldsymbol{x})U^{-1}\right) < 0, \tag{4.103}$$

then the system is asymptotically stable w.r.t $\boldsymbol{x} = 0$ over Λ, thus Λ is contained in the entire region of attraction.

Proof Let $V(\boldsymbol{x}) = \boldsymbol{x}^T U^T U \boldsymbol{x}$ be a Lyapunov function candidate. Since U is nonsingular, $V(\boldsymbol{x})$ is positive definite. Differentiating $V(\boldsymbol{x})$ yields

$$\begin{aligned}\dot{V}(x) &= \dot{\boldsymbol{x}}^T U^T U \boldsymbol{x}\\ &= \boldsymbol{x}^T A(\boldsymbol{x})^T U^T U\boldsymbol{x} + \boldsymbol{x}^T U^T U A(\boldsymbol{x})\boldsymbol{x}\\ &= \boldsymbol{x}^T U^T \left((U^{-1})^T A(\boldsymbol{x})^T U^T + UA(\boldsymbol{x})U^{-1}\right) U\boldsymbol{x}.\end{aligned} \tag{4.104}$$

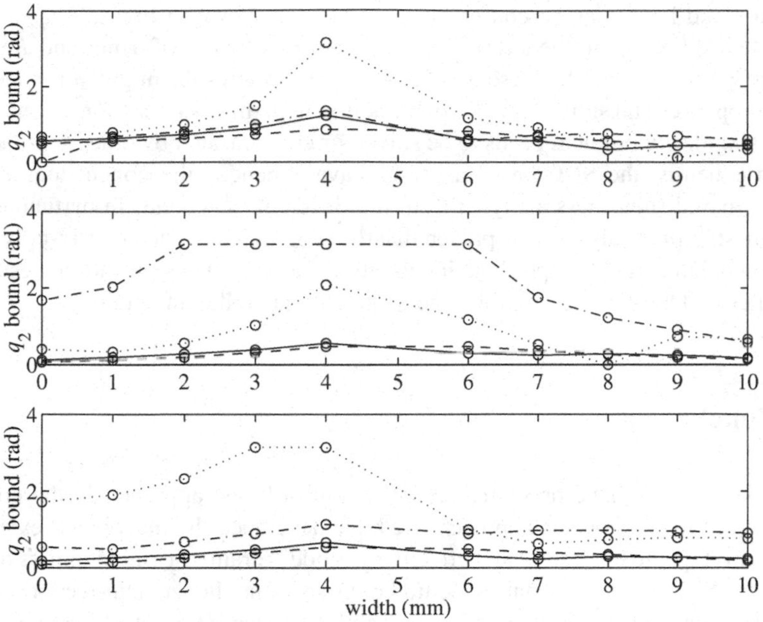

Fig. 4.23 Results of robust stability analyses based on Theorem 4.5; $\bar{q}_2$ of the region of attraction versus payload width; *Top* single-frequency oscillation, *Middle* double-frequency oscillation, *Bottom* Bretschneider oscillation; *solid line* K_{1j}, *dashed line* K_{2j}, *dashed-dotted line* K_{1bjred}, *dotted line* K_{2bjred}

Then by (4.103) and U's nonsingularity $\dot{V}(\boldsymbol{x})$ is negative definite over Λ. Hence the proof is complete.

Based on the theorem, we calculated each Λ for each controller, that is, a subset of the region of attraction that satisfies the condition (4.103) for a given payload variation. The resulting Λ can be represented as

$$\boldsymbol{q} \times \dot{\boldsymbol{q}} \times \boldsymbol{\xi} = [-\pi/10, \pi/10] \times [-\bar{q}_2, \bar{q}_2] \times [-5, 5] \times [-12, 12] \times \mathbb{R}^m, \quad (4.105)$$

where $\bar{q}_2$ is the bound for q_2 and varies with the payload variations, which is displayed in Fig. 4.23. In Fig. 4.23, each graph is associated with each oscillation case, and compares the respective controllers. As in the case of maximum $Re(\lambda_i(A(\boldsymbol{x})))$, the profiles with K_{1j} and K_{2j} via γ-iteration are almost the same, and quite similar regardless of the oscillation cases. On the other hand, those with the controllers via D–K iteration vary depending on the oscillation cases, however, reveal apparently larger bounds than those with K_{1j}'s and K_{2j}'s, which show the stronger robustness.

This robust stability analysis method based on Theorem 4.5 is more stringent than the above μ-based one, because any approximation on the evaluated system is not used. This method is essentially based on the concept of "quadratic stability," and is very easy to implement and useful. However, the result provides only a sufficient condition and hence can be conservative. In fact, as shown in Fig. 4.23, $\bar{q}_2$'s were calculated to be zero in the cases of K_{1b1red} with the width of 0 mm and K_{2b2red} with the width of 8 mm. In the sense of continuity, those results might not reflect the actual properties and similar regions of attraction with those of near conditions. This sort of singularity problem needs to be solved or compensated by other approaches.

Consequently, the SDC-based analyses have enhanced the conclusions of the above μ-based robustness analyses from the different viewpoint. In particular, the stringent stability analyses have proven that the proposed extended matrix-polytope-based model uncertainty representation is useful for control design and the resulting controller will have promising robustness against the model uncertainties.

4.5 Conclusions

In this section, we have presented the $\mathcal{H}_\infty$-control-based approach to the control problem of OBMs. For this approach, we have developed the machinery, extended matrix polytope, to efficiently and effectively model parametric uncertainties due to the payload variation. The control method consists of nonlinear state-feedback control to reduce the nonlinearity of the system and the linear $\mathcal{H}_\infty$ control scheme. In the control design process, how to model uncertainties and what kinds of weighting functions to be used for $\mathcal{H}_\infty$ control design are important. Therefore, we have designed four types of controllers with differences in uncertainty model and/or weighting functions, and compared their properties by examining their frequency responses, poles, and zeros and by performing robustness analysis with two types of tests. One of the robustness analysis is the constant-scaling μ-analysis to accommodate the time-varying uncertainties and the other one is based on the SDC form and the Lyapunov theory. From all the results and arguments, we have concluded as:

- All the $\mathcal{H}_\infty$ controllers have the corresponding frequency modes of the base oscillation as their poles, thus according to the internal model principle they are expected to effectively reduce the disturbance due to the base oscillation;
- The weighting function W_n for sensor noise plays an important role in not only noise reduction but also robustness enhancement;
- The extended matrix polytope can efficiently and effectively represent parametric uncertainties due to the payload variation, and can provide stronger robustness to the resultant controllers.

Again it should be noted that those results are not exclusive for the OBM control problems but extensively applicable to wider class of mechanical system control problems.

Chapter 5
Simulations and Experiments for the $\mathcal{H}_\infty$-Control-Based Approach

Abstract This chapter presents control performance evaluations of the $\mathcal{H}_\infty$-control-based approach by simulations and hardware experiments separately from Chap. 4. We have performed demonstrations of control performance considering the three types of control problems and three patterns of base oscillations mentioned in Chap. 4, and the four types of $\mathcal{H}_\infty$-controllers presented in Chap. 4. For those demonstration cases, we investigate the respective control performance, with respect to nominal-case performance, robust control performance against the payload variations, influence of sensor error, and moreover comparison with the conventional PID control. Those results show that the $\mathcal{H}_\infty$-control-based approach and the extended matrix polytope are practically useful and effective from the viewpoint of control performance.

5.1 Introduction

This chapter presents simulation and experimental evaluations on the $\mathcal{H}_\infty$-control-based approach presented in the previous chapter in terms of the control design method, robustness analyses of the obtained control systems. As demonstration cases, the three types of control problems, i.e., attitude control in local coordinates, attitude control in global coordinates, position control in global coordinates, and the three types of base oscillation, i.e., the single-frequency oscillation, the double-frequency one, the Bretschneider one, are considered comparing the four types of $\mathcal{H}_\infty$ controllers.

This chapter is organized as follows. First, in Sect. 5.2 we present nominal-case control performances, where examination of sensor error influence and effectiveness of the weighting function W_n, and comparison with the conventional PID control are included. Then, Sect. 5.3 demonstrates robust control performances. The last section gives some conclusions.

M. Toda, *Robust Motion Control of Oscillatory-Base Manipulators*,
Lecture Notes in Control and Information Sciences 463,
DOI 10.1007/978-3-319-21780-2_5

5.2 Simulations and Experiments (Nominal-Case Performance)

First, we focus on nominal-case performances of the proposed approach comparing the four types of controller K_{1j}'s, K_{1bjred}'s, K_{2j}'s, and K_{2bjred} mentioned in the previous chapter. Further, we will compare those controllers with a PID controller to evaluate our approach fairly. We conducted simulations with respect to all the combinations of three types of oscillation (single-frequency, double-frequency, and Bretschneider oscillations), three types of control problems (attitude control in local coordinates, attitude control in global coordinates, and position control in global coordinates), and the four types of controller. Further, choosing some cases among them, we investigated influence of sensor errors, comparison with the conventional PID control, and carried out experiments using the experimental apparatus introduced in Chap. 3. In all the demonstration cases presented in this section, the nominal case, i.e., with the payload width of 5 mm is considered

5.2.1 Base Oscillation

Figure 5.1 shows the profiles of three types of base oscillation employed for demonstrations, where the top graph depicts the single-frequency oscillation, the second does the double-frequency one, and the bottom does the Bretschneider oscillation. In

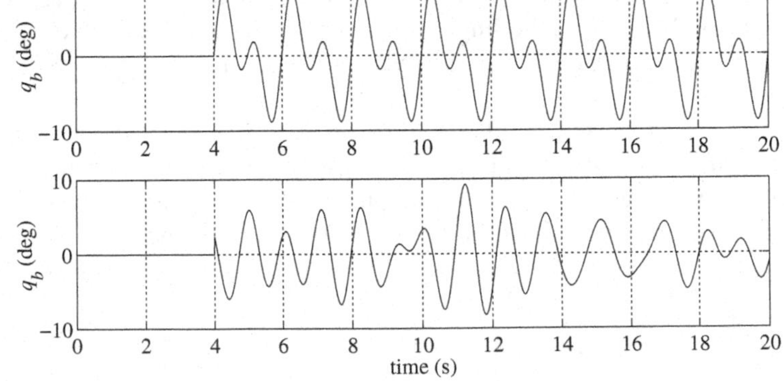

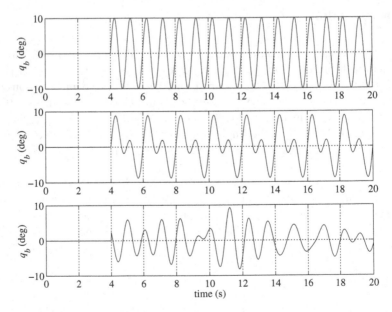

Fig. 5.1 Base motion for simulations and experiments; *top* the single-frequency oscillation, *middle* the double-frequency oscillation, *bottom* the Bretschneider oscillation

attitude control cases, the base starts oscillation at 4 s, while in position control case, it starts at 10 s. In every case, the whole period of demonstration is 20 s. It should be noted that the power of oscillation per frequency of the single-frequency case is the largest one being compared with the others, which will be reflected by control simulation results.

5.2.2 Global Coordinates

Here we define some variables for demonstrations. r_1 and r_2 denote the reference signals to the respective manipulator joints, while r_{g1} and r_{g2} are the reference ones in the global coordinates: $r_{g1} = q_b + r_1, r_{g2} = q_b + q_1 + r_2$. Similarly, $q_{g1} = q_b + q_1$ and $q_{g2} = q_b + q_1 + q_2$ are the angles of the links in the global coordinates.

5.2.3 PID Control

For comparison with the proposed control approach, the conventional PID control is also demonstrated. The adopted PID controllers to the tracking error $\boldsymbol{e}$ are as in the following:

$$\tau_1 = F_{P1}e_1 + F_{D1}\dot{e}_1 + F_{I1}\int_0^t e_1 d\tau \tag{5.1}$$

$$\tau_2 = F_{P2}e_2 + F_{D2}\dot{e}_2 + F_{I2}\int_0^t e_2 d\tau \tag{5.2}$$

where $\boldsymbol{\tau} = [\tau_1, \tau_2]^T$ is the control torque vector, F_{Pi}, F_{Di}, and F_{Ii} are the proportional, derivative, and integral control gains for Link i respectively. As mentioned below, those gains are determined by using the optimization scheme with a criterion of root mean square control error by simulation similarly to the $\mathcal{H}_\infty$ control case. The optimization scheme for PID control is based on a quadratic programming algorithm over [0.1, 10], [0.1, 10], [0.01, 1] for F_{I1}, F_{P1}, F_{D1} and [0.01, 1], [0.01, 1], [0.001, 0.1] for F_{I2}, F_{P2}, F_{D2} respectively.

5.2.4 Attitude Control

For attitude control, we demonstrate two types of problems, i.e., in local coordinates and in global coordinates respectively. The initial conditions are $q_1(0) = q_2(0) = q_b(0) = 0\,\text{deg}$, $\dot{q}_1(0) = \dot{q}_2(0) = \dot{q}_b(0) = 0\,\text{deg/s}$. The whole duration is 20 s. As displayed in Fig. 5.1, the base starts oscillation at 4 s with the respective oscillations.

In the local coordinate demonstration, the references r_1 and r_2 are step function such that $r_1 = 0 \to 30\,\text{deg}$ and $r_2 = 0 \to 60\,\text{deg}$ at 10 s. In the global coordinate one, the references r_{g1} and r_{g2} are step functions such that $r_{g1} = 0 \to 30\,\text{deg}$ and $r_{g2} = 0 \to 90\,\text{deg}$ at 10 s, that is, $r_1 = -q_b \to -q_b + 30$ and $r_2 = -q_1 - q_b \to -q_1 - q_b + 90\,\text{deg}$. To avoid overshoot, we applied a first-order low-pass filter with the time constant 0.8 to those step signals.

As mentioned above, to optimize the linear state-feedback gains $F = [F_1, F_2]$, the 401 time series angle tracking error vectors of simulation

$$\boldsymbol{e}_s = \begin{bmatrix} e_1(16.00), e_1(16.01), \ldots, e_1(20.00) \\ e_2(16.00), e_2(16.01), \ldots, e_2(20.00) \end{bmatrix} \tag{5.3}$$

are adopted as the steady errors and the $||\boldsymbol{e}_s||_F$ is calculated as the criterion.

Further, we defined the corresponding root mean square error (RMSE) as $\bar{e} = ||\boldsymbol{e}_s||_F/\sqrt{802}$, which aims at making us understand control performances intuitively.

By repeating simulations of the case of attitude control in global coordinates with double-frequency oscillation under the above conditions, F for the $\mathcal{H}_\infty$ controller and PID gains the PID controller were optimized. To determine F, the $\mathcal{H}_\infty$ controller K_{22} was employed, and then the obtained F is commonly used for all the $\mathcal{H}_\infty$ controllers. As a result of the optimization,

- $F_1 = 66$ and $F_2 = 24$ were obtained;
- the optimal PID controller gains were as $F_{I1} = 2.9$, $F_{P1} = 2.3$, $F_{D1} = 0.05$, $F_{I2} = 0.19$, $F_{P2} = 0.32$, $F_{D2} = 0.01$.

5.2.5 Position Control

Next, we demonstrate position control with respect to the center of the payload in the global coordinates. With a contrast to attitude control, position control requires an online solution to the inverse kinematics problem. Let (r_x, r_y) denote the reference position of the center of the payload in the global coordinates. Then, the joint references can be obtained as follows:

$$\begin{aligned} r'_x &= \sqrt{L_x^2 + L_y^2} \cos\left\{q_b + \tan^{-1}\left(\frac{L_y}{L_x}\right)\right\} - r_x \\ r'_y &= r_y - \sqrt{L_x^2 + L_y^2} \sin\left\{q_b + \tan^{-1}\left(\frac{L_y}{L_x}\right)\right\} \\ \psi &= \frac{r'^2_x + r'^2_y - l_1^2 - l_2^2}{2l_1 l_2} \end{aligned} \tag{5.4}$$

$$r_2 = \tan^{-1}\left(\frac{\sqrt{1-\psi^2}}{\psi}\right)$$

$$r_1 = \tan^{-1}\left(\frac{r'_x}{r'_y}\right) - \tan^{-1}\left(\frac{l_2 \sin r_2}{l_1 + l_2 \cos r_2}\right) - q_b. \tag{5.5}$$

The initial conditions are $q_1(0) = q_2(0) = q_b(0) = 0\,\text{deg}$, $\dot{q}_1(0) = \dot{q}_2(0) = \dot{q}_b(0) = 0\,\text{deg/s}$, which correspond to $(r_x(0), r_y(0)) = (0.10, 0.20)\,(\text{m})$. The whole duration is 20 s. (r_x, r_y) are two step reference signals such that $(0.10, 0.20) \rightarrow (0.0826, 0.1988)$ at 4 s, which position corresponds to $q_1(0) = q_b(0) = 0$ and $q_2(0) = 5\,\text{deg}$, and then $(0.0826, 0.1988) \rightarrow (0.06, 0.13)$ at 6 s, and the base starts oscillation at 10 s, which are intended to avoid multiple solutions of the inverse kinematics. To avoid overshoot, we applied the same low-pass filter as in the attitude control case.

For controller parameter optimization, the 401 time series tracking "distance" errors as $\boldsymbol{e}_s = [e(16.00), e(16.01), \ldots, e(20.00)]$ are adopted by conducting simulations with the double-frequency oscillation. The corresponding mean of the steady-state error was defined as $\bar{e} = ||\boldsymbol{e}_s||_F/\sqrt{401}$.

5.2.6 Simulation Results

Table 5.1 shows the results of RMSE $\bar{e}$ in the respective control simulations. First, let us focus on the differences among the base oscillation cases. As pointed out before, $\bar{e}$'s in the case of single-frequency are larger those in the other cases, which is due to the oscillation power per frequency. Then, we compare the types of problems. $\bar{e}$'s of attitude control in global coordinates are larger than those of the local coordinate

Table 5.1 Results of RMSE $\bar{e}$ in the respective control simulations

	K_{11}	K_{1b1red}	K_{21}	K_{2b1red}
Attitude-l (deg)	0.1081	0.1488	0.1047	0.1250
Attitude-g (deg)	0.2396	0.3463	0.2152	0.3309
Position (mm)	0.4080	0.8215	0.3972	0.6125
	K_{12}	K_{1b2red}	K_{22}	K_{2b2red}
Attitude-l (deg)	0.1033	0.1236	0.1072	0.1047
Attitude-g (deg)	0.1723	0.2097	0.1361	0.1501
Position (mm)	0.2285	0.3443	0.2276	0.2521
	K_{13}	K_{1b3red}	K_{23}	K_{2b3red}
Attitude-l (deg)	0.1059	0.1116	0.1085	0.1002
Attitude-g (deg)	0.1916	0.1955	0.1486	0.1424
Position (mm)	0.2393	0.4508	0.2358	0.3213

Attitude-l attitude control in local coordinates, *attitude-g* attitude control in global coordinates, *position* position control

case. In the global coordinate problems, not only disturbance suppression but also tracking control to oscillatory trajectory is required, which makes the problem more difficult than the local coordinate ones.

Now we compare the types of controllers. Regarding the point of explicit model uncertainty consideration, that is, comparing K_{ij} and K_{ibjred}, K_{ij} performs better than K_{ibjred}. This result is natural since the demonstration is based on the nominal case. Next let us investigate effect of the weighting function W_n for sensor noise. As those results suggests, W_n is very effective in those demonstrations, which is because the joint angle sensor resolution of 0.18 deg is rather coarse. This point will be later discussed in more detail. In almost all the cases, K_{2j}'s reveal the best performances, while K_{1bjred}'s do the worst performances. K_{1j}'s and K_{2bjred}'s are in the middle class. However, taking into account of the sensor resolution of 0.18 deg, those controllers exhibit successful performances, which has proven that our proposed approach is considerably effective and useful.

Next, in order to investigate time series data, we present the simulation results of the proposed $\mathcal{H}_\infty$ controllers in the cases of the single-frequency oscillation, which can demonstrate the essential features of the four types of controllers and the three types of control problems, because the oscillation power per frequency is the largest as pointed out above. Figures 5.2, 5.3, 5.4, 5.5, 5.6, 5.7, 5.8, 5.9, 5.10, 5.11, 5.12, and 5.13 show the time series data of the control simulation results, respectively, in the

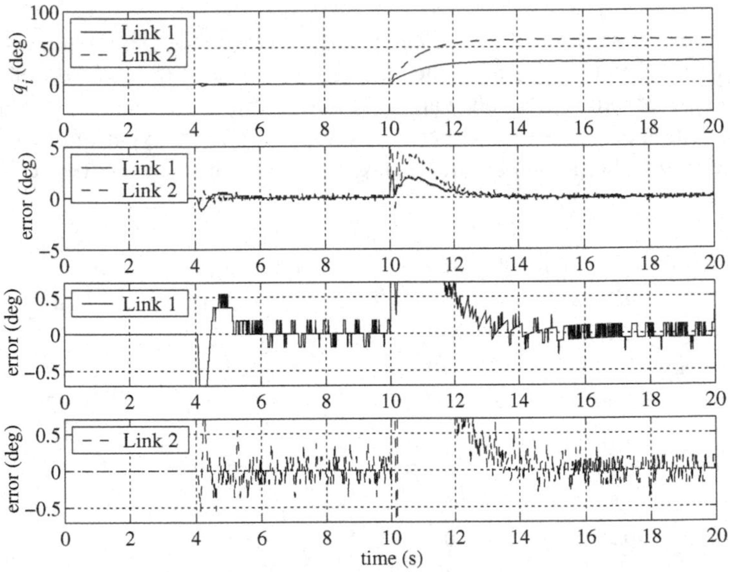

Fig. 5.2 Simulation results of attitude control in local coordinates with the controller K_{11} for the single-frequency oscillation; *top* the attitudes of the links in the local coordinates, second from the *top* the tracking errors on a normal scale, third from the *top* the tracking error of Link 1 on a fine scale, *bottom* the tracking error of Link 2 on a fine scale

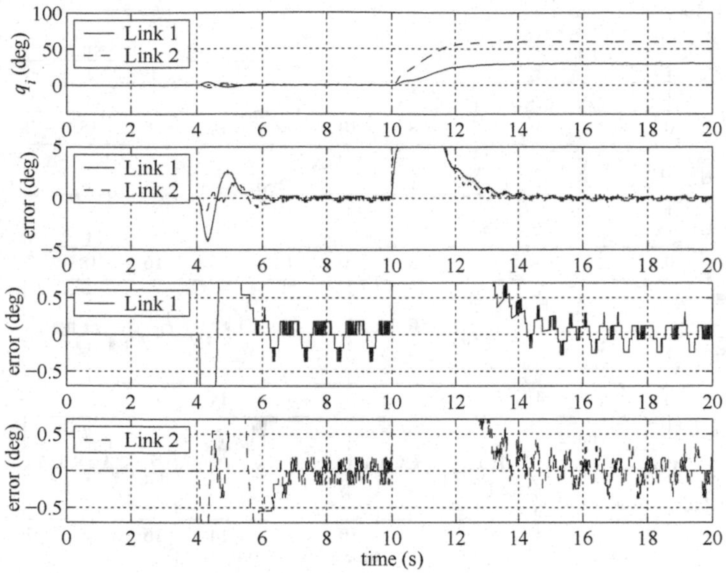

Fig. 5.3 Simulation results of attitude control in local coordinates with the controller K_{1b1red} for the single-frequency oscillation; *top* the attitudes of the links in the local coordinates, second from the *top* the tracking errors on a normal scale, third from the *top* the tracking error of Link 1 on a fine scale, *bottom* the tracking error of Link 2 on a fine scale

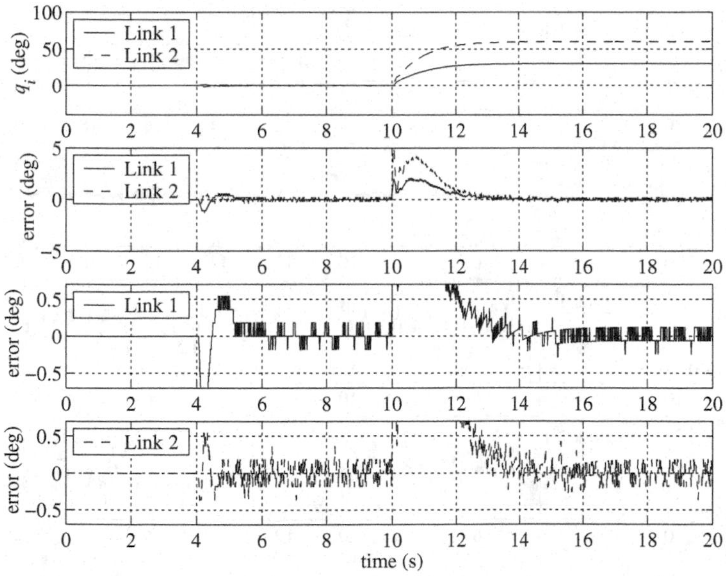

Fig. 5.4 Simulation results of attitude control in local coordinates with the controller K_{21} for the single-frequency oscillation; *top* the attitudes of the links in the local coordinates, second from the *top* the tracking errors on a normal scale, third from the *top* the tracking error of Link 1 on a fine scale, *bottom* the tracking error of Link 2 on a fine scale

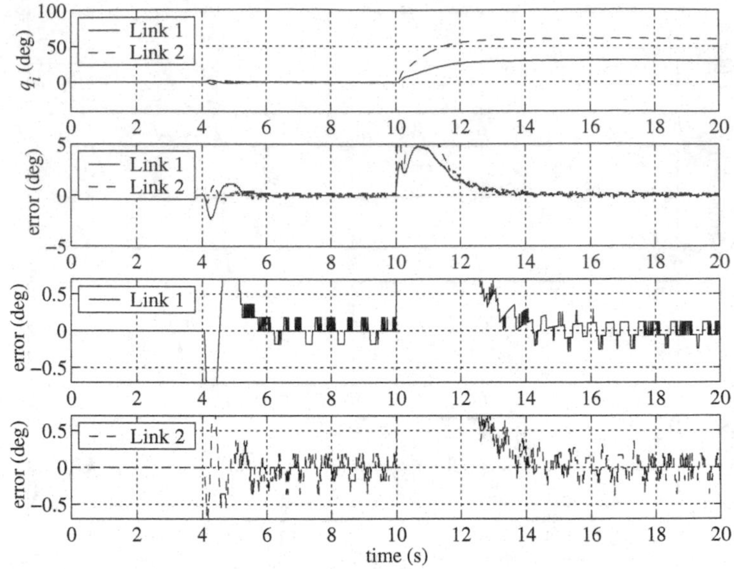

Fig. 5.5 Simulation results of attitude control in local coordinates with the controller K_{2b1red} for the single-frequency oscillation; *top* the attitudes of the links in the local coordinates, second from the *top* the tracking errors on a normal scale, third from the *top* the tracking error of Link 1 on a fine scale, *bottom* the tracking error of Link 2 on a fine scale

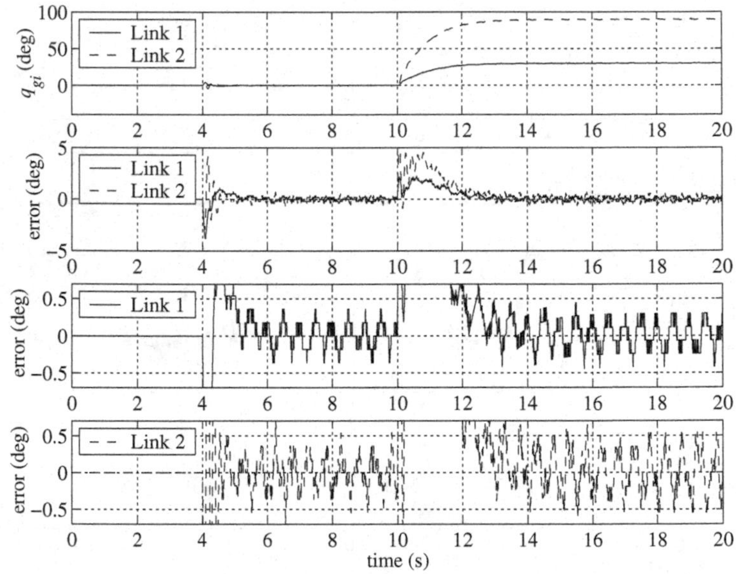

Fig. 5.6 Simulation results of attitude control in global coordinates with the controller K_{11} for the single-frequency oscillation; *top* the attitudes of the links in the global coordinates, second from the *top* the tracking errors on a normal scale, third from the *top* the tracking error of Link 1 on a fine scale, *bottom* the tracking error of Link 2 on a fine scale

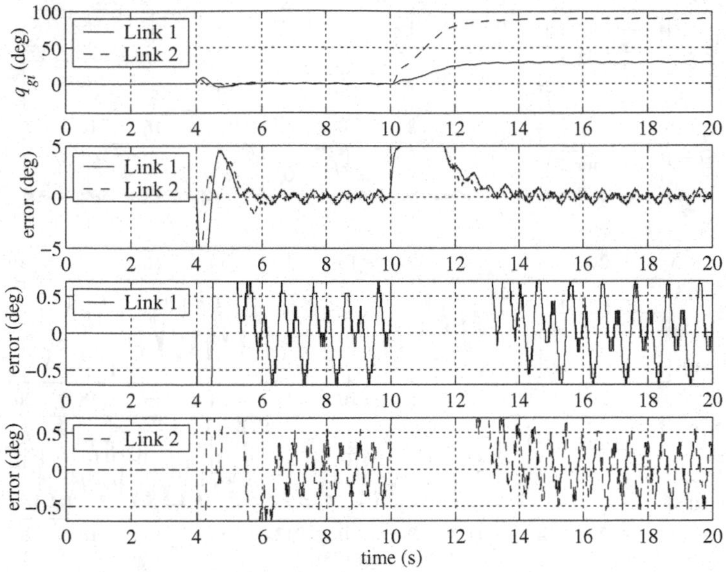

Fig. 5.7 Simulation results of attitude control in global coordinates with the controller K_{1b1red} for the single-frequency oscillation; *top* the attitudes of the links in the global coordinates, second from the *top* the tracking errors on a normal scale, third from the *top* the tracking error of Link 1 on a fine scale, *bottom* the tracking error of Link 2 on a fine scale

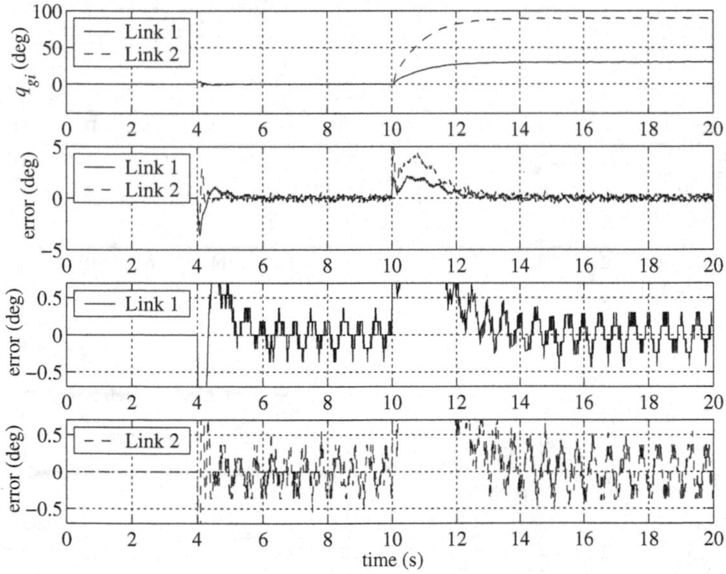

Fig. 5.8 Simulation results of attitude control in global coordinates with the controller K_{21} for the single-frequency oscillation; *top* the attitudes of the links in the global coordinates, second from the *top* the tracking errors on a normal scale, third from the *top* the tracking error of Link 1 on a fine scale, *bottom* the tracking error of Link 2 on a fine scale

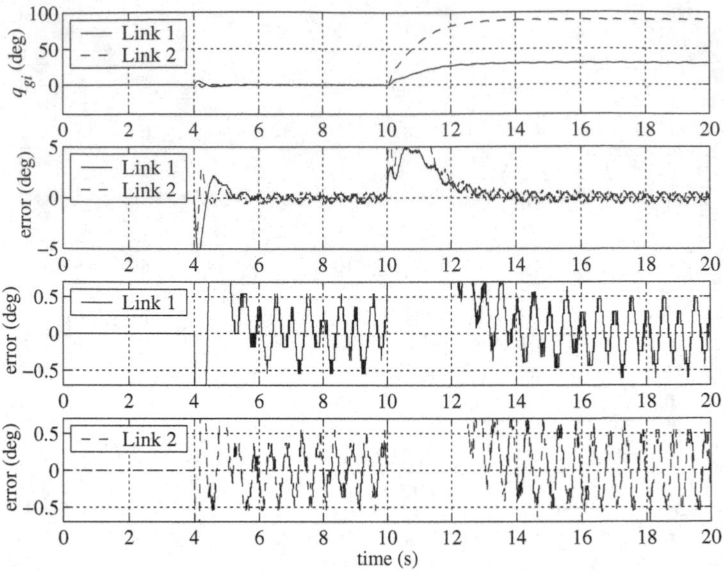

Fig. 5.9 Simulation results of attitude control in global coordinates with the controller K_{2b1red} for the single-frequency oscillation; *top* the attitudes of the links in the global coordinates, second from the *top* the tracking errors on a normal scale, third from the *top* the tracking error of Link 1 on a fine scale, *bottom* the tracking error of Link 2 on a fine scale

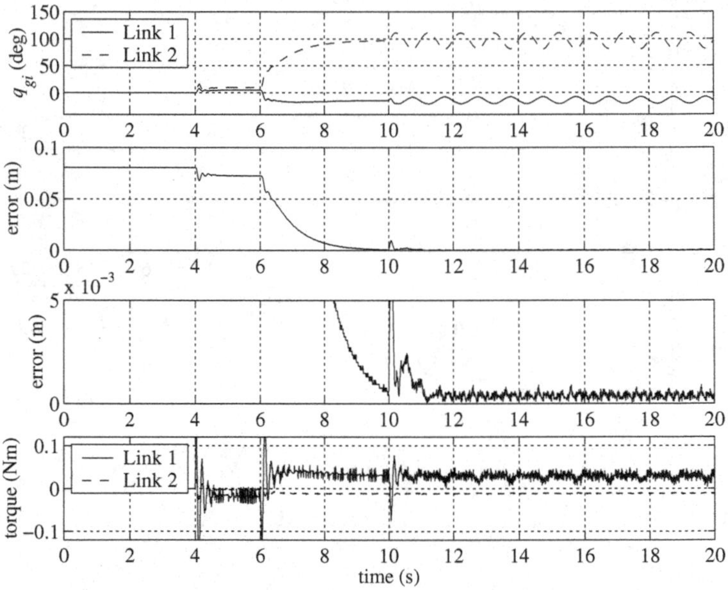

Fig. 5.10 Position-control simulation results with the controller K_{11} for the single-frequency oscillation; *top* the attitudes of the links in the global coordinates, second from the *top* the distance errors from the final goal on a normal scale, third from the *top* the distance errors from the final goal on a fine scale, *bottom* the control torques

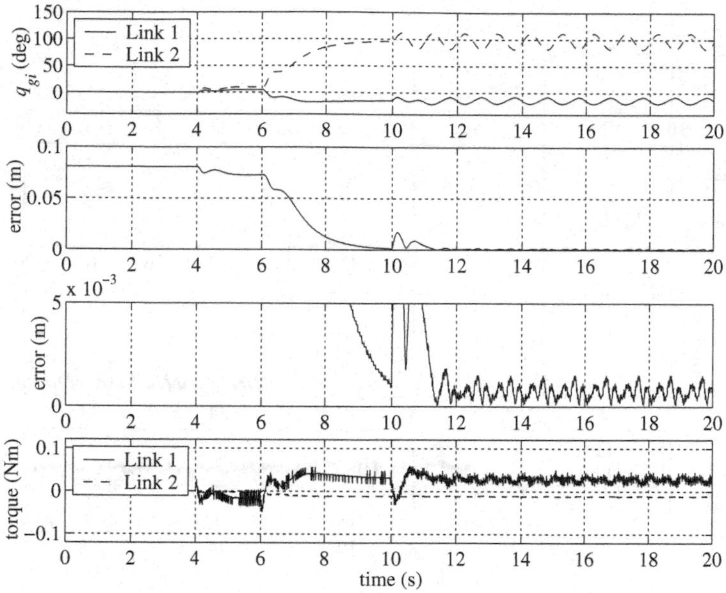

Fig. 5.11 Position control simulation results with the controller K_{1b1red} for the single-frequency oscillation; *top* the attitudes of the links in the global coordinates, second from the *top* the distance errors from the final goal on a normal scale, third from the *top* the distance errors from the final goal on a fine scale, *bottom* the control torques

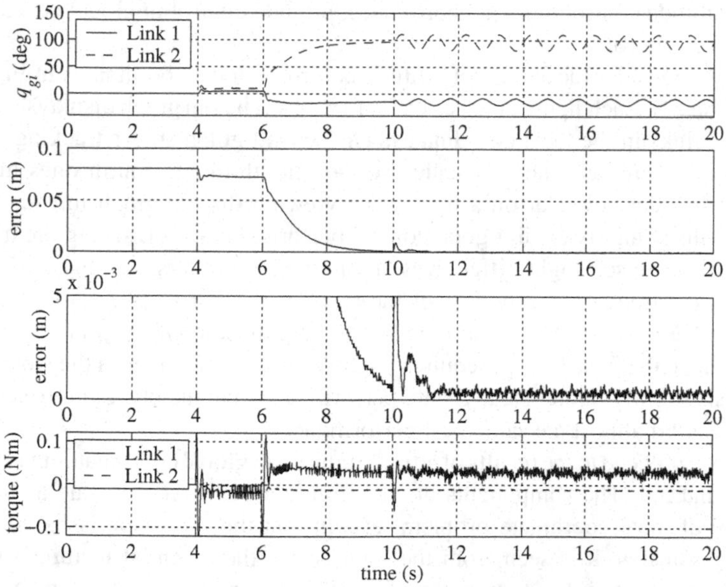

Fig. 5.12 Position control simulation results with the controller K_{21} for the single-frequency oscillation; *top* the attitudes of the links in the global coordinates, second from the *top* the distance errors from the final goal on a normal scale, third from the *top* the distance errors from the final goal on a fine scale, *bottom* the control torques

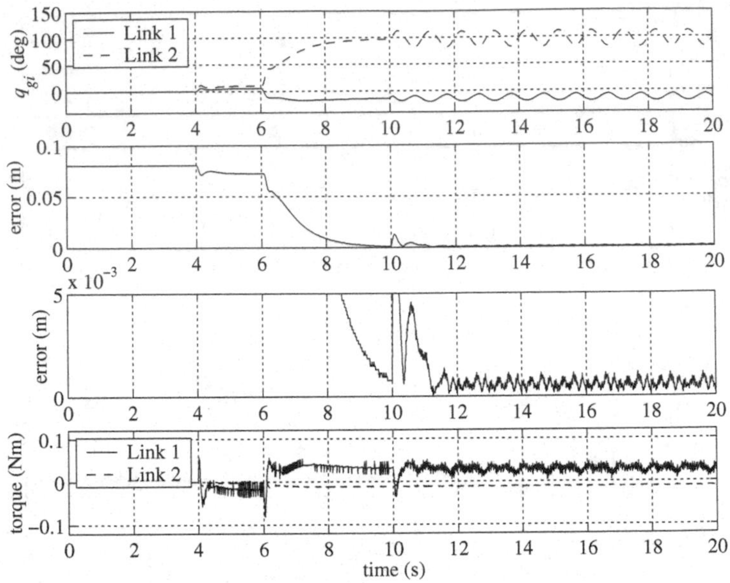

Fig. 5.13 Position control simulation results with the controller K_{2b1red} for the single-frequency oscillation; *top* the attitudes of the links in the global coordinates, second from the *top* the distance errors from the final goal on a normal scale, third from the *top* the distance errors from the final goal on a fine scale, *bottom* the control torques

order of attitude control in local coordinates, attitude control in global coordinates, and position control.

First, we consider the results of attitude control in local coordinates in Figs. 5.2, 5.3, 5.4, and 5.5. Each figure consists of four graphs. The top graph displays attitudes of the two links in the local coordinates. The second graph shows tracking control errors of the joints on a normal scale. Further, the third and fourth ones magnify the respective control error on a fine scale. Viewing the top graph in each figure, all the results similarly exhibit good control performances. But, seeing the tracking error profiles, we see slight differences among the controllers. See Fig. 4.18 which represents the frequency responses of the respective controllers K_{11}–K_{2b1red}, and recall that the gains of K_{1b1red} are the lowest and those of K_{2b1red} are the second lowest. Due to the fact, K_{1b1red} requires slightly longer time to reach the steady state and reveals slightly poorer steady-state performance than the other controllers. K_{11} and K_{21} exhibit almost the same best performances.

Then, we proceed to the results of attitude control in global coordinates in Figs. 5.6, 5.7, 5.8, and 5.9. The components of each figure are almost the same as those of the figure of local coordinate problem, except that the top graph is based on the global coordinates. It is seen from the figures that the essential features in terms of differences among the four types of controllers are the same as in the above arguments, and are even more manifest. This fact is also suggested by the data of $\bar{e}$'s in Table 5.1.

Next, we present the results of position control in global coordinates in Figs. 5.10, 5.11, 5.12, and 5.13. In each figure, the top graph displays the attitude of the two links in the global coordinates; the second represents the distance error from the final goal position; the third one magnifies the distance error on a fine scale; and the bottom graph does the control torques of the respective joints. Again, these results reveal the same features in terms of differences among the controllers.

Finally, we see the simulation results of control torques. Figures 5.14 and 5.15 show the control torques of the joints 1 and 2, respectively, with K_{11}. Note that the profiles of control torques are essentially the same in the cases of the other controllers. As in the order of control problems in the figures, control efforts are most required in the attitude control in local coordinates, and the attitude control in global coordinates is the second. Regarding the differences in the power of control torques and the differences in the resulting control performances, it seems that in the control problem where the less power of control torques are required, the gap between the resultant control performances of K_{ij} and K_{ibjred} is the larger.

Consequently, by investigating all the simulation results carefully, we have presented the fundamental properties of the proposed $\mathcal{H}_\infty$ controllers and their relationships in the three types of control problems. As long as the nominal situation is concerned, all the proposed controllers work effectively, in particular the controller K_{2j}'s perform best.

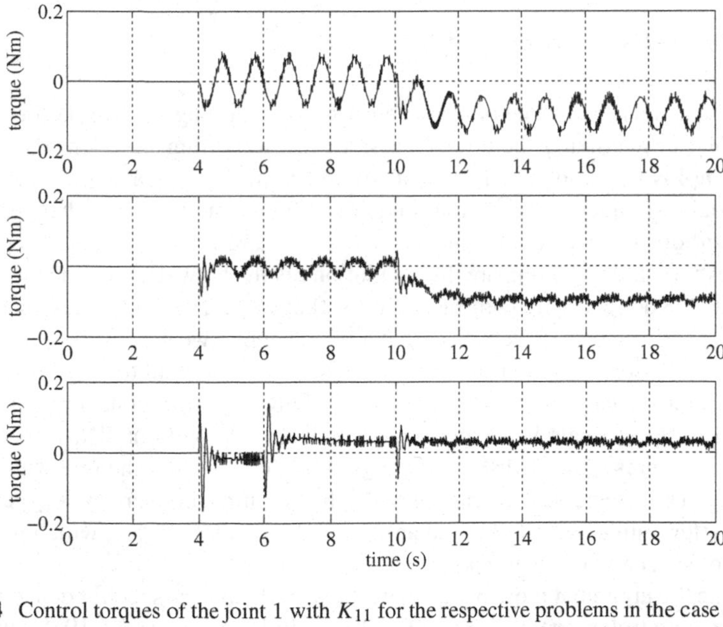

Fig. 5.14 Control torques of the joint 1 with K_{11} for the respective problems in the case of single-frequency oscillation; *top* attitude control in local coordinates, *middle* attitude control in global coordinates, *bottom* position control in global coordinates

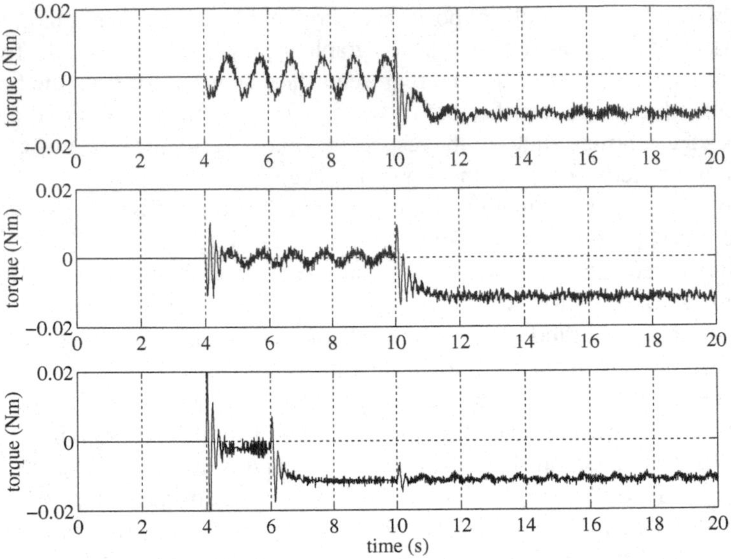

Fig. 5.15 Control torques of the joint 2 with K_{11} for the respective problems in the case of single-frequency oscillation; *top* attitude control in local coordinates, *middle* attitude control in global coordinates, *bottom* position control in global coordinates

5.2.7 Influence of Sensor Error

We here investigate influence of sensor error to the resulting control performances, in particular emphasizing the effectiveness of the weighting function W_n. To this end, we employ the simulation results of attitude control in global coordinates in the case of double-frequency oscillation. In order to focus on differences between with W_n and without W_n, we compare the results of K_{12} and K_{22}.

Figures 5.16 and 5.17 compare the simulation results between with the low sensor resolution 0.18 deg and a high sensor resolution 0.0018 deg, i.e., 100 times finer, in the case of K_{12} and K_{22} respectively. In each figure, the upper two graphs show the results with the low sensor resolution and the lower ones do those with the high resolution. First, consider the case of the low resolution, where K_{22} with consideration of W_n in control design reveals better performances than K_{12} without W_n, in particular for Link 2. The resulting RMSE $\bar{e}$ with K_{22} is 0.1361 deg, and the one with K_{12} is 0.1723 deg. Then, regrading the case of the high resolution, conversely K_{12} performs slightly better with $\bar{e} = 0.0636$ deg than K_{22} with $\bar{e} = 0.0675$ deg, which might be difficult to be seen in these graphs.

To make the argument more obvious, we present the power spectral density of the steady-state control errors (time = 16.00–20.00) in Figs. 5.18 and 5.19. Figure 5.18 compare those situations for Link1, and Fig. 5.19 for Link 2. First, focus on the peaks at very low frequencies, that is, 0.5 and 1 Hz, which have appeared in all the graphs

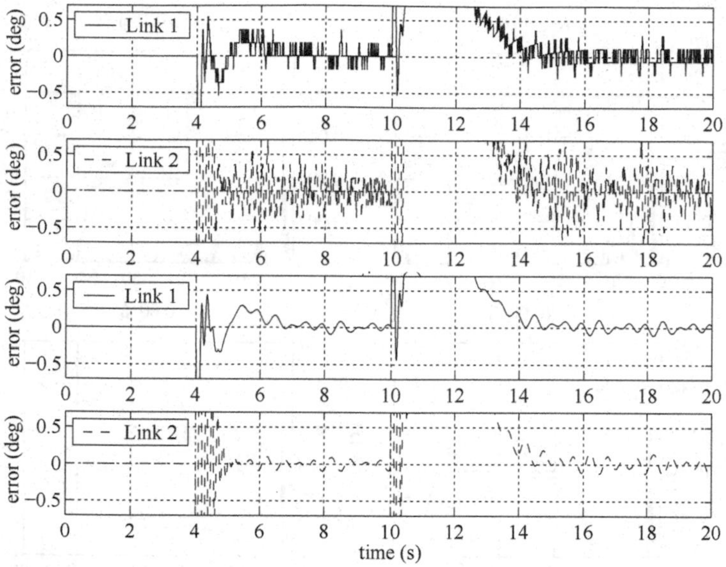

Fig. 5.16 Comparison of sensor resolution in attitude control in global coordinates with K_{12} in the case of double-frequency oscillation; the *upper* two graphs: simulation results with the sensor resolution of 0.18 deg, the *lower* two graphs: simulation results with the sensor resolution of 0.0018 deg

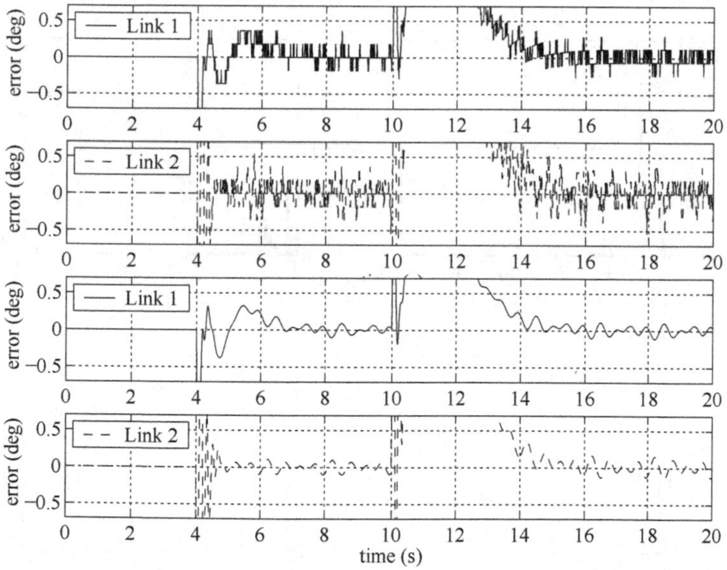

Fig. 5.17 Comparison of sensor resolution in attitude control in global coordinates with K_{22} in the case of double-frequency oscillation; the *upper* two graphs: simulation results with the sensor resolution of 0.18 deg, the *lower* two graphs: simulation results with the sensor resolution of 0.0018 deg

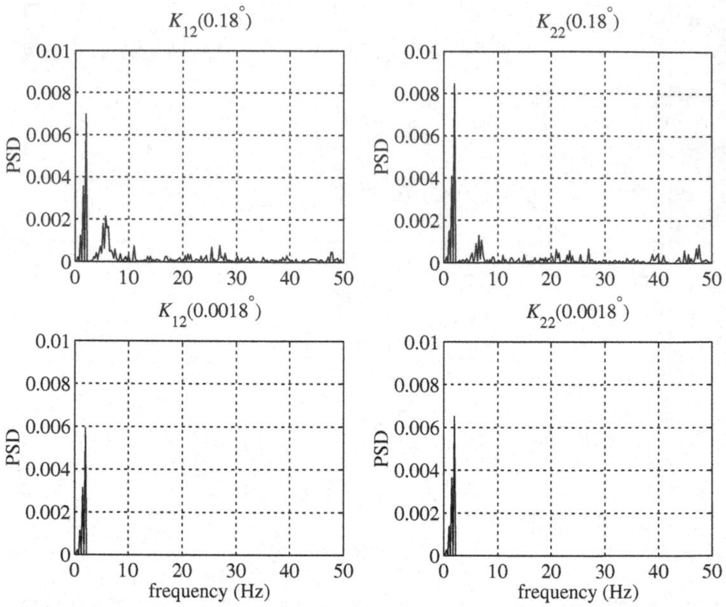

Fig. 5.18 Power spectral density (PSD) of the steady-state control errors (time = 16.00–20.00) of the joint 1

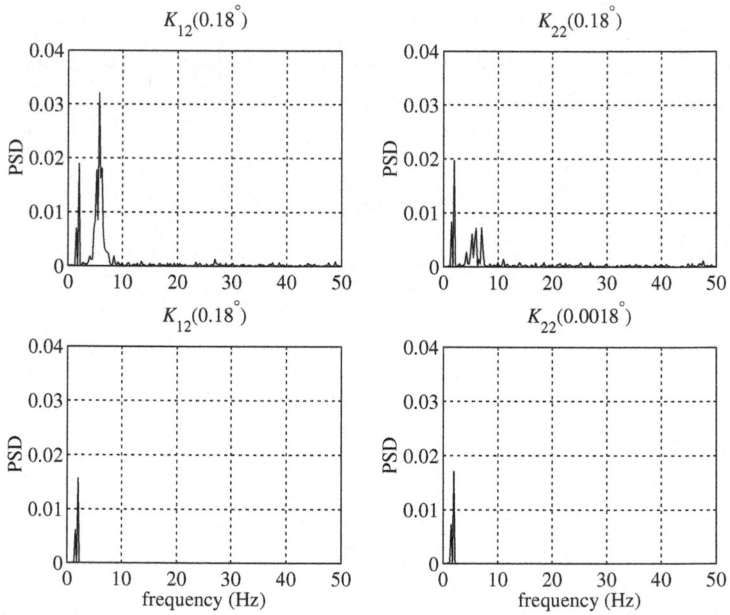

Fig. 5.19 Power spectral density (PSD) of the steady-state control errors (time = 16.00–20.00) of the joint 2

and represents the residual errors due to the base oscillation. Interestingly enough, in the case of the low resolution (see the upper graphs), although the RMSE $\bar{e}$ with K_{22} is less than that with K_{12}, its component due to the base oscillation with K_{22} is greater than that with K_{12}. Further, it is seen that K_{12} suffers from more high-frequency noises due to sensor errors than K_{22}. Then, in the case of the high resolution (see the lower graphs), the components due to sensor errors have disappeared, and K_{12} exhibits less power of control error than K_{22}.

From the above argument, we see that the weighting function W_n for sensor noise in $\mathcal{H}_\infty$ control design plays an important role in such a case the sensor noise is a problem, however, slightly sacrifices the tracking control performance of the control system.

5.2.8 Experimental Results and Comparison with PID Control

Focusing on attitude control in global coordinates with the double-frequency oscillation, we compare the simulation results and experimental ones, and further compare the proposed approach with the conventional PID control. For this demonstration, the $\mathcal{H}_\infty$ controller K_{22}, which reveals the best performance in the nominal case, has been chosen. And it should be noted that the PID gains have been optimized for this control problem as described above.

First, we compare the simulation and experimental results with K_{22} as shown in Figs. 5.20 and 5.21 respectively. These figures have the same structure as the corresponding ones in the case of single-frequency oscillation in the previous section. Then, it is seen that not only the simulation results but also the experimental ones exhibit successful control performances and both the profiles in the simulation and experimental results are considerably consistent, although the experimental control errors for Link 2 are slightly noisier than the simulation ones. The RMSE $\bar{e}$ of simulation is 0.1361 deg, while $\bar{e}$ of experiment is 0.1842 deg. Additionally, these results have proven that the obtained parameters in the dynamical model of the experimental apparatus are accurate and the apparatus are appropriately developed.

Next, we present the simulation and experimental results with the PID controller. Figures 5.22 and 5.23 display the simulation and experimental results with the PID controller respectively. Concentrating on the steady-state control errors after the start of step tracking control at 10 s in the simulation results in Fig. 5.22, the PID controller performs successfully and better with $\bar{e} = 0.1142$ deg than K_{22} with $\bar{e} = 0.1361$ deg. However, unfavorable residual oscillations due to the base oscillation have appeared on the control errors from 4 to 10 s for both the links. This difference has come from the difference in the attitudes in the both cases, and suggests that the performance of the PID controller might largely vary depending on the manipulator attitudes. Recall that the PID gains have been tuned by taking only $\bar{e}$ into account. Moreover, let us see the experimental results in Fig. 5.23. Then, it is seen that Link 2 suffers from tremendous large oscillation, which implies that the proportional and/or derivative gains of the PID controller are too large under the experimental environments.

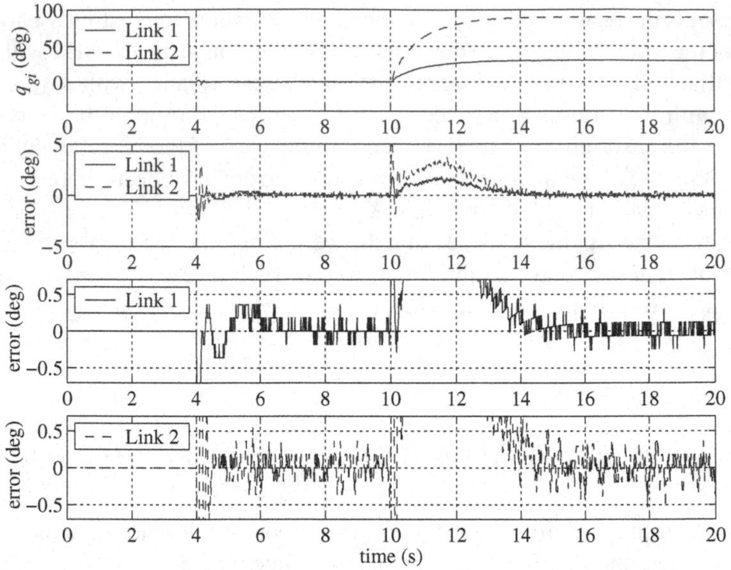

Fig. 5.20 Simulation results of attitude control in global coordinates with the controller K_{22} for the double-frequency oscillation; *top* the attitudes of the links in the global coordinates, second from the *top* the tracking errors on a normal scale, third from the *top* the tracking error of Link 1 on a fine scale, *bottom* the tracking error of Link 2 on a fine scale

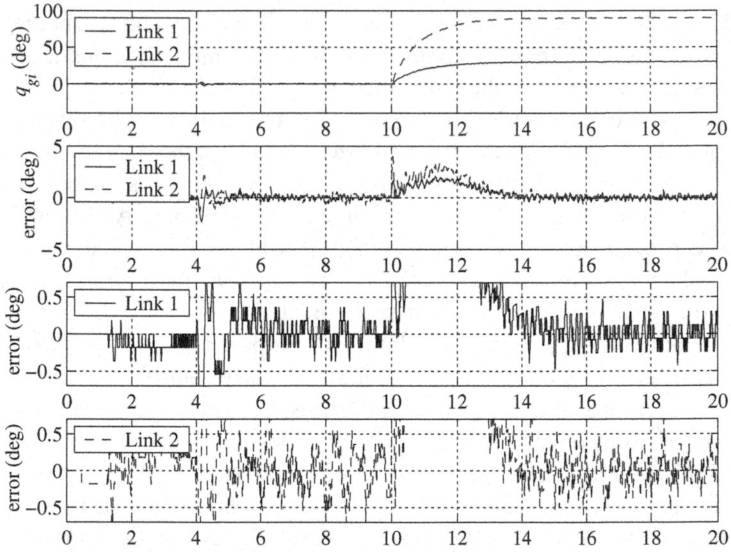

Fig. 5.21 Experimental results of attitude control in global coordinates with the controller K_{22} for the double-frequency oscillation; *top* the attitudes of the links in the global coordinates, second from the *top* the tracking errors on a normal scale, third from the *top* the tracking error of Link 1 on a fine scale, *bottom* the tracking error of Link 2 on a fine scale

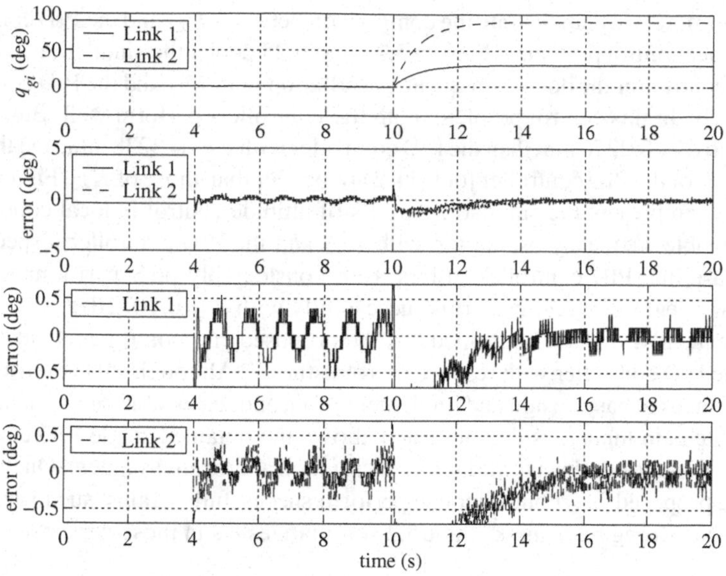

Fig. 5.22 Simulation results of attitude control in global coordinates with the PID controller for the double-frequency oscillation; *top* the attitudes of the links in the global coordinates, second from the *top* the tracking errors on a normal scale, third from the *top* the tracking error of Link 1 on a fine scale, *bottom* the tracking error of Link 2 on a fine scale

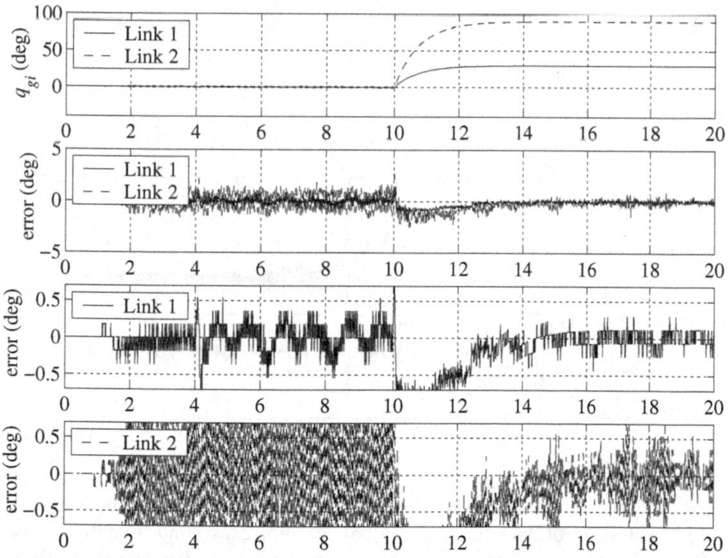

Fig. 5.23 Experimental results of attitude control in global coordinates with the PID controller for the double-frequency oscillation; *top* the attitudes of the links in the global coordinates, second from the *top* the tracking errors on a normal scale, third from the *top* the tracking error of Link 1 on a fine scale, *bottom* the tracking error of Link 2 on a fine scale

Furthermore, we demonstrate the comparison between K_{22} and the PID controller for the other control problems. Figures 5.24 and 5.25 depict the simulation results of position control for the double-frequency oscillation with K_{22} and the PID controller respectively. In this control problem, both the controllers perform well. But, K_{22} is better with $\bar{e} = 0.2276$ mm than the PID controller with $\bar{e} = 0.3238$ mm, and the control torques of the PID controller for Link 2 are noisier than those of K_{22}. Figures 5.26 and 5.27 then present the simulation results of attitude control in local coordinates for the double-frequency oscillation with K_{22} and the PID controller respectively. In this case, the PID controller suffers from considerably poor performance with $\bar{e} = 0.3857$ deg, while K_{22} performs successfully with $\bar{e} = 0.1072$ deg.

Consequently, we have confirmed that the proposed $\mathcal{H}_\infty$ controller is much superior to the PID controller with respect to "robustness." Although PID control enjoys the advantages of easy design and implementation, and reasonable performance with appropriate gain tuning, as demonstrated above, the obtained single PID controller cannot be flexible or robust to variations of situation and/or environment. On the other hand. the proposed $\mathcal{H}_\infty$ controller can perform successfully against such variations, and thus has strong robustness. Table 5.2 summarizes $\bar{e}$'s in those demonstrations.

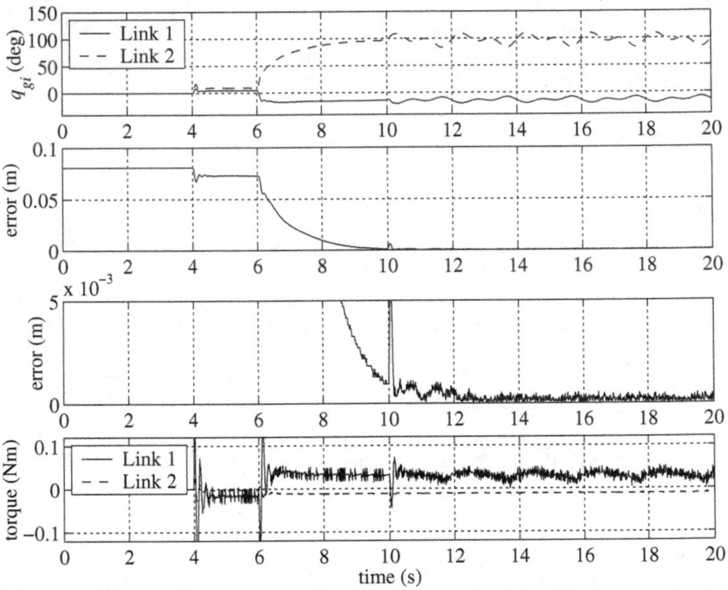

Fig. 5.24 Position control simulation results with the controller K_{22} for the double-frequency oscillation; *top* the attitudes of the links in the global coordinates, second from the *top* the distance errors from the final goal on a normal scale, third from the *top* the distance errors from the final goal on a fine scale, *bottom* the control torques

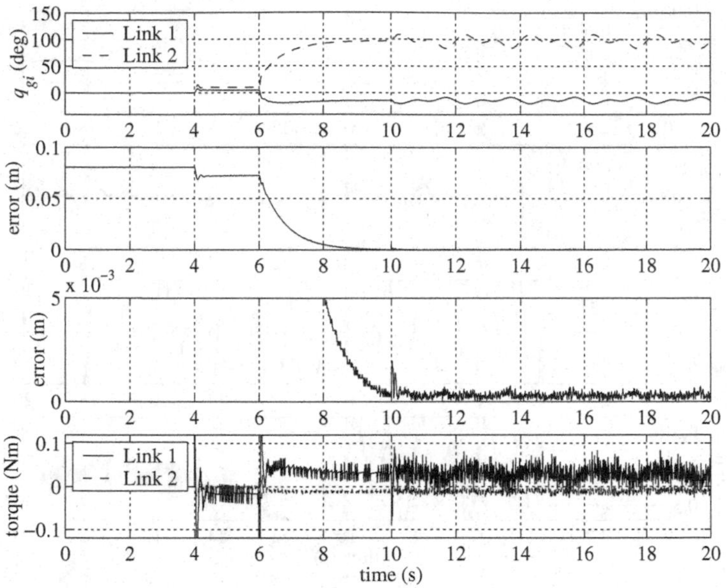

Fig. 5.25 Position control simulation results with the PID controller for the double-frequency oscillation; *top* the attitudes of the links in the global coordinates, second from the *top* the distance errors from the final goal on a normal scale, third from the *top* the distance errors from the final goal on a fine scale, *bottom* the control torques

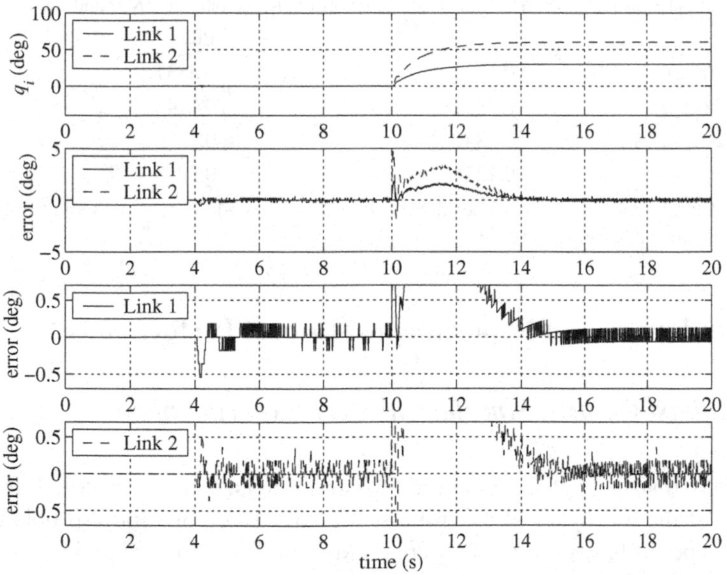

Fig. 5.26 Simulation results of attitude control in local coordinates with the controller K_{22} for the double-frequency oscillation; *top* the attitudes of the links in the local coordinates, second from the *top* the tracking errors on a normal scale, third from the *top* the tracking error of Link 1 on a fine scale, *bottom* the tracking error of Link 2 on a fine scale

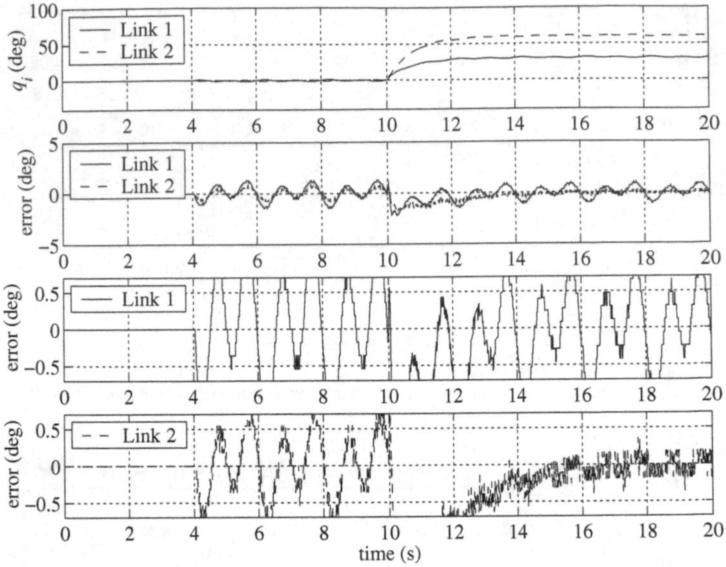

Fig. 5.27 Simulation results of attitude control in local coordinates with the PID controller for the double-frequency oscillation; *top* the attitudes of the links in the local coordinates, second from the *top* the tracking errors on a normal scale, third from the *top* the tracking error of Link 1 on a fine scale, *bottom* the tracking error of Link 2 on a fine scale

Table 5.2 Results of RMSE $\bar{e}$ in demonstrations for comparison between K_{22} and the PID control

	K_{22}	PID
Attitude-l (deg)	0.1072	0.3857
Attitude-g (deg)	0.1361	0.1142
Attitude-g-ex (deg)	0.1842	0.2303
Position (mm)	0.2276	0.3238

Attitude-l attitude control in local coordinates, *attitude-g* attitude control in global coordinates (-ex denotes the experimental results)

5.3 Simulations and Experiments (Robust Performance)

5.3.1 Robust Control Simulations and Experiments

To evaluate robustness of the proposed $\mathcal{H}_\infty$ controllers, simulations for all the combinations of the four types of controllers, the three types of control problems, and the three types of base oscillations were conducted. In addition to them, robust control experiments for selected cases among them were also conducted in comparison with the PID controller. In order to implement physical parametric variations of the manipulator, all the demonstrations deployed payload exchanges. As described in Chap. 3, the payload of the experimental OBM is exchangeable with its width of

0, 1, 2, ..., 10 mm (See Table 3.4 again to understand how those width variations induce physical parametric variations).

Except the payload, the simulation and experimental methods are the same as those of the nominal-case ones in Sect. 5.2. For robust control experiments, being based on the attitude control in global coordinates for the double-frequency oscillation, the $\mathcal{H}_\infty$ controllers K_{22} and K_{2b2red}, and the PID controller were employed. One of the objectives of the experiments is to investigate effectiveness of the explicit consideration of model uncertainties in control design, that is, to see how robust K_{2b2red} can be in the experimental environment.

5.3.2 Results

The results of the robust control demonstrations are mainly represented and evaluated by using RMSE $\bar{e}$'s. Figures 5.28, 5.29, and 5.30 show graphs of $\bar{e}$ versus payload width in the respective oscillation cases. Each figure consists of the graph of attitude control in local coordinates (top), the one of attitude control in global coordinates (middle), and the one of position control in global coordinates (bottom).

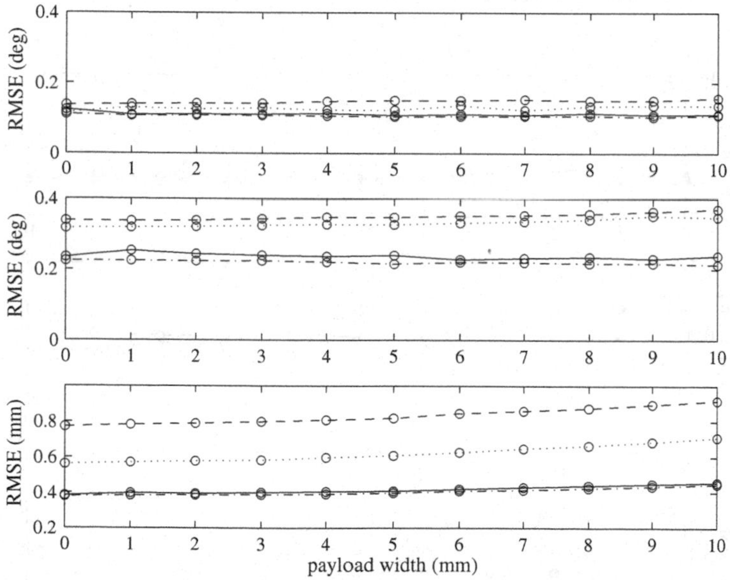

Fig. 5.28 RMSE $\bar{e}$ versus payload width of the simulation results in the case of single-frequency oscillation; *top* attitude control in local coordinates, *middle* attitude control in global coordinates, *bottom* position control in global coordinates; *solid* K_{11}, *dashed* K_{1b1red}, *dashed-dotted* K_{21}, *dotted* K_{2b1red}

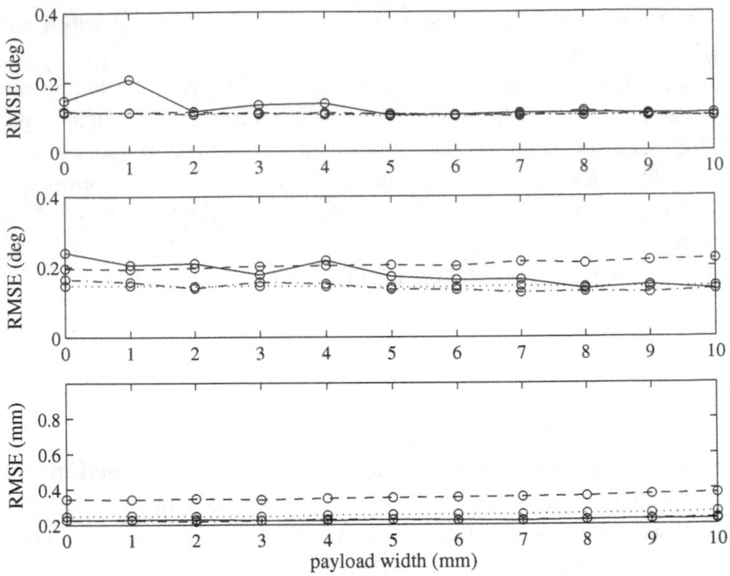

Fig. 5.29 RMSE $\bar{e}$ versus payload width of the simulation results in the case of double-frequency oscillation; *top* attitude control in local coordinates, *middle* attitude control in global coordinates, *bottom* position control in global coordinates; *solid* K_{12}, *dashed* K_{1b2red}, *dashed-dotted* K_{22}, *dotted* K_{2b2red}

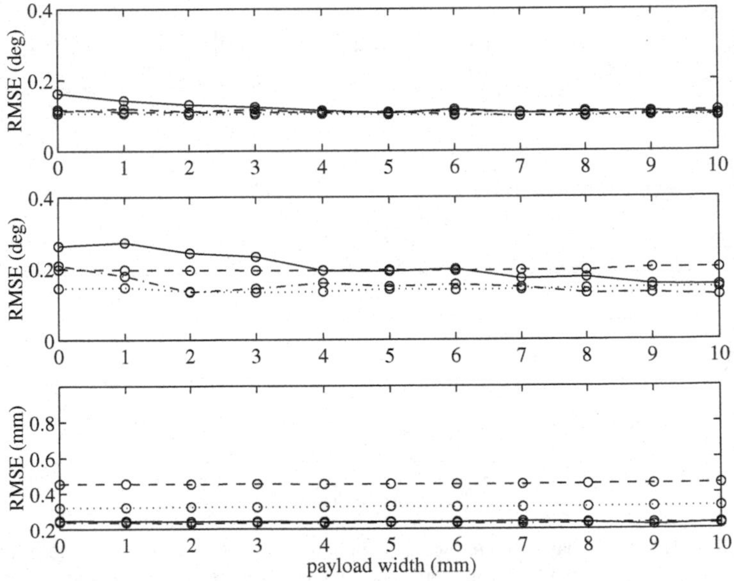

Fig. 5.30 RMSE $\bar{e}$ versus payload width of the simulation results in the case of Bretschneider oscillation; *top* attitude control in local coordinates, *middle* attitude control in global coordinates, *bottom* position control in global coordinates; *solid* K_{13}, *dashed* K_{1b3red}, *dashed-dotted* K_{23}, *dotted* K_{2b3red}

Here we should emphasize that, in the nominal case, the essential features and relationships of the four types of controllers do not vary with the oscillation cases, however, which is not the case in the robust control demonstrations. First, let us see Fig. 5.28 representing the single-frequency oscillation case, where $\bar{e}$'s of every controller are invariant over the payload widths and thus the relationships among the controllers are the same as those in the nominal case. Then, carefully investigating Figs. 5.29 and 5.30, we see that the relationships of the controllers are different from those in Fig. 5.28 and in the nominal case, in the attitude control cases and with the smaller widths. In particular, in the graphs of attitude control in global coordinates in both the cases of double-frequency and Bretschneider oscillations, K_{ibjred} reveals better performance than the corresponding K_{ij}, which show effectiveness of the explicit consideration of model uncertainties in control design by using extended matrix polytopes. This effectiveness becomes the more apparent when the oscillation pattern becomes the more complicated. K_{2b2red} and K_{2b3red} have strong robustness and reveal the best worst-case performances in such situations.

On the other hand, one question has arisen to us. Considering the results of the position control in global coordinates, the relationships of all the controllers are completely invariant against such model variations in all the cases of base oscillations, although the essential control schemes required are similar with those for the attitude control in global coordinates. Then, where does such difference come from? At the moment, we have not got a clear answer to the question. However, since a stark contrast between the two control problems is the difference in the power of required control torques as displayed in Figs. 5.14 and 5.15, we suppose this point lead to the difference in the relationships of the controllers. If the force applied to the mass is the larger, variation in the mass leads to the larger change in the motion, hence to keep the motion in such a case, the more robustness is necessary for the controller.

Next, we address the experimental results. Figure 5.31 shows RMSE $\bar{e}$ versus payload width of the experimental results of attitude control in global coordinates in the case of double-frequency oscillation similarly with those of simulations, but includes the results of the PID controller. In the experiments, the payload widths of 0, 2, 5, 8, 10 mm were adopted. Each trial was repeated three times, and the mean of $\bar{e}$'s was obtained as the final datum. As seen from the figure, K_{22} and K_{2b2red} reveal strong robustness even in the experimental situations, while the performance of the PID controller dramatically deteriorates at width of 2 mm. As shown later, the experimental OBM with the PID control under such a condition exhibited unfavorable chattering phenomenon. Hence we avoided conducting the trial with the width of 0 mm for the PID controller.

Figures 5.32, 5.33, and 5.34 display the control error profiles of Link 2 on a fine scale over the payload width variations with K_{22}, K_{2b2red}, and the PID controller respectively. Figures 5.32 and 5.33 show that both K_{22} and K_{2b2} can provide strongly robust and successful performances over such payload variations. However, by carefully checking those figures, in Fig. 5.32 the profiles at widths of 8 and 10 mm are slightly different from those at widths of 2 and 5 mm, whereas in Fig. 5.33 all the profiles are almost the same. Thus, K_{2b2red} is more robust than K_{22}. On the other hand,

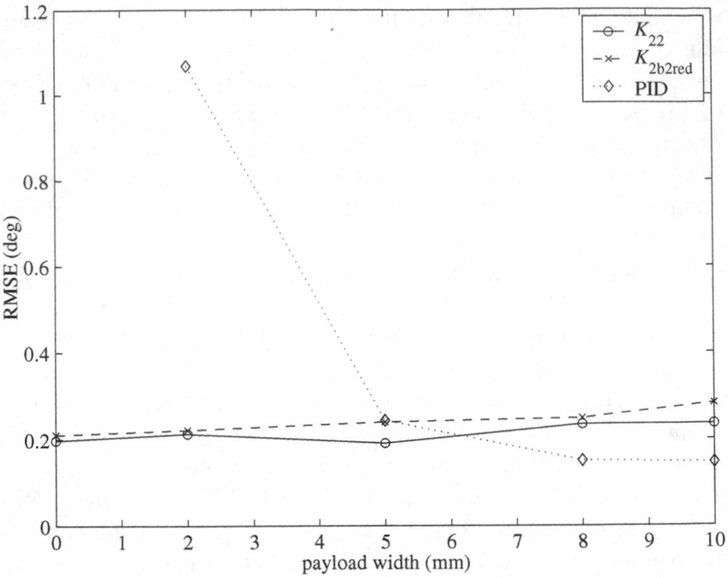

Fig. 5.31 RMSE $\bar{e}$ versus payload width of the experimental results of attitude control in global coordinates in the case of double-frequency oscillation; *solid* K_{22}, *dashed* $K_{2b21red}$, *dotted* PID

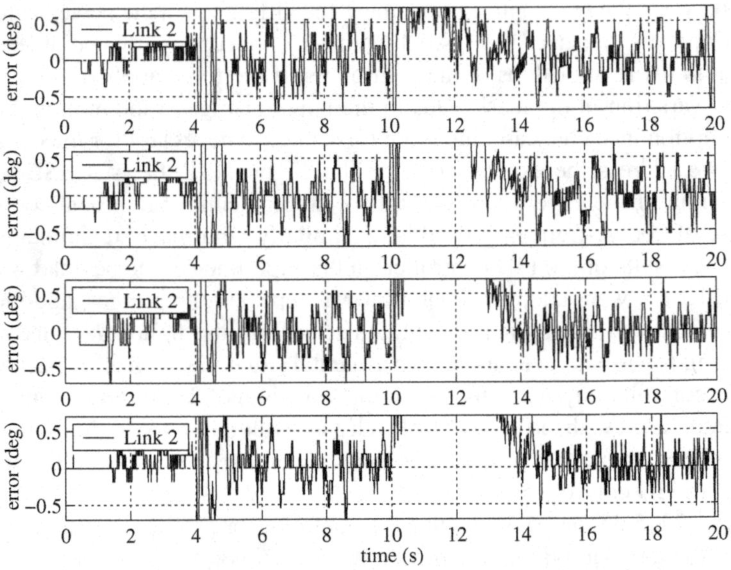

Fig. 5.32 Experimental results of attitude control in global coordinates for Link 2 with K_{22} over the payload width variations; *top* 10 mm, second: 8 mm, third: 5 mm, *bottom* 2 mm

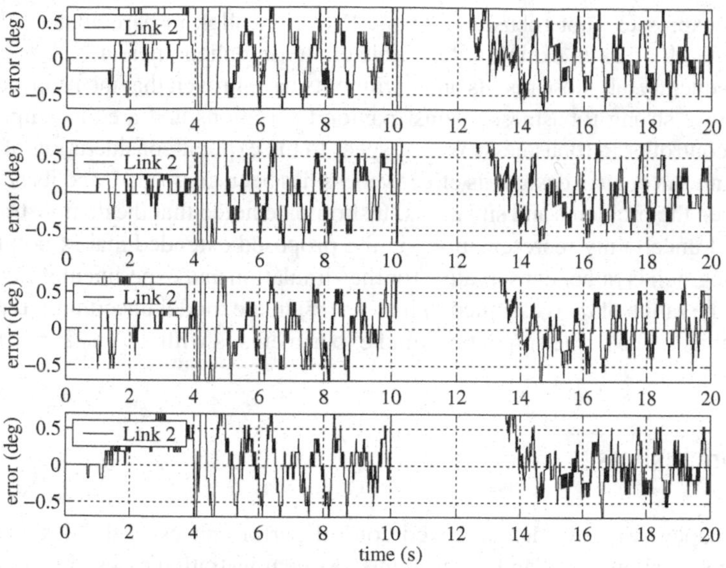

Fig. 5.33 Experimental results of attitude control in global coordinates for Link 2 with K_{2b2red} over the payload width variations; *top* 10 mm, second: 8 mm, third: 5 mm, *bottom* 2 mm

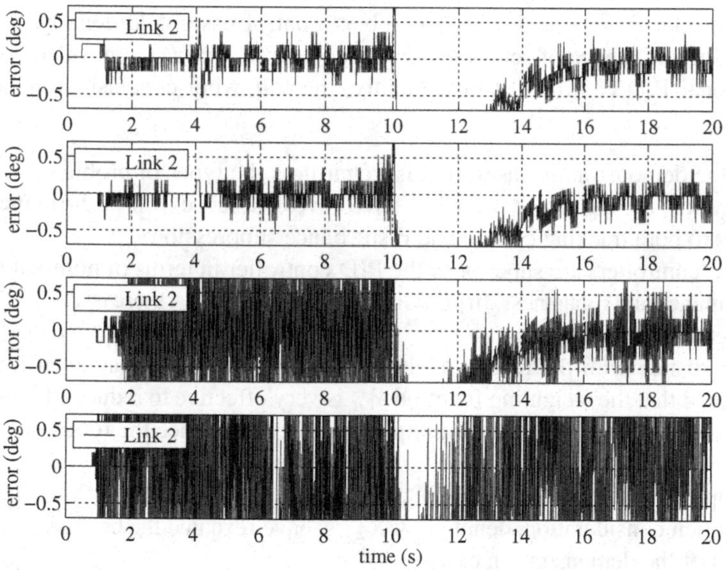

Fig. 5.34 Experimental results of attitude control in global coordinates for Link 2 with the PID controller over the payload width variations; *top* 10 mm, second: 8 mm, third: 5 mm, *bottom* 2 mm

Fig. 5.34 expresses poor robustness of the PID controller, in particular the bottom graph at width of 2 mm exhibit unfavorable chattering phenomenon.

Consequently, all those results and arguments have proven that the proposed controllers have strong robustness against payload variations and are also superior to the PID controller in robustness. With respect to the explicit consideration of model uncertainties in control design, its effectiveness has been confirmed, and its necessity depends on the problems and situations. It should be noted that the demonstration of the model uncertainty representation via the proposed extended matrix polytope in this monograph is rather conservative for the simulation and experimental examples, because the range of q_2 is assumed to be $[-\pi, \pi]$ in the model uncertainty representation but its range in the simulations and experiments is within $[-\pi/2, \pi/2]$.

5.4 Conclusions

In this chapter, we have demonstrated control performances of the designed $\mathcal{H}_\infty$ controllers by simulations and experiments. As demonstration cases, the three types of control problems and the three types of base oscillations have been considered. In the respective demonstration cases, we have compared the four types of $\mathcal{H}_\infty$ controllers, i.e., with W_n or not, and with explicit consideration of model uncertainties or not, in both the nominal case and robust control case. Additionally, we have examined the influence of sensor error and compared the $\mathcal{H}_\infty$ controllers with the conventional PID controller. Consequently, the following conclusions have been obtained:

- In all the demonstration cases, that is, regardless the types of problems and base oscillations, all the $\mathcal{H}_\infty$ controllers have revealed successful performance with respect to both tracking control and disturbance suppression;
- the $\mathcal{H}_\infty$ controllers are superior to the PID controller in terms of nominal control performance and robustness. In particular, in the robustness, there is a stark contrast between those controllers;
- In accordance with the results of analyses in the previous chapter, it has been confirmed that the weighting function W_n is very effective to reduce the influence of sensor error, and also plays an important role to enhance the robustness of the control system;
- In comparison of the four types of $\mathcal{H}_\infty$ controllers, the controller with W_n and without explicit consideration, denoted by K_{2j}'s, have revealed the best performances in most of the demonstration cases;
- With respect to the effectiveness of explicit consideration of model uncertainties via extended matrix polytope, in robust control cases, the corresponding controllers have exhibited strong robustness. The control performances of the controllers without W_n are poorer than those of K_{2j}'s, however, by adding W_n the performances can be greatly improved to be near to those of K_{2j}'s. Furthermore, in some robust control cases, these controllers have the best performances. Thus, those results have

convinced us that the extended matrix polytope is effective not only in robustness analysis but also in control design.

It should be noted that the oscillatory base problems obviously contains ordinary problems, i.e., with non-oscillatory bases, as a special class. Therefore, the proposed methodology with some tools such as an extended polytope is extensively applicable to general control problems of mechanical systems.

Chapter 6
Motion Control Using a Sliding-Mode-Control-Based Approach

Abstract In this chapter, we present the other highlight of the monograph, the sliding-mode control (SMC)-based approach for the OBM robust control problems. In our attempt to apply the SMC framework to such problems, we have developed a novel nonlinear sliding surface, which is called the "rotating sliding surface with variable-gain integral control (RSSI)." The advantageous feature of this method is the control system can achieve successful tracking control and disturbance rejection in not only steady state but also transient state with less control inputs. We present the control design method, stability analyses on the RSSI and control performance demonstrations by simulations in comparison with those of the $\mathcal{H}_\infty$-control-based approach. The results show that, in the ideal situation in terms of sensor resolution, sampling period, control input limitation, the SMC-RSSI system is considerably superior to the $\mathcal{H}_\infty$ controllers, however once such ideal conditions have been violated, its performance are dramatically deteriorated. Therefore, the SMC-based approach is promising, but needs to be improved in the sense of practical implementation.

6.1 Introduction

In this chapter, we present the other proposed control design approach, that is, a sliding mode control (SMC)-based one. Theoretically, SMC can be expected to achieve good robustness and control performance without *a priori* knowledge on frequencies of the base oscillation [5, 20, 39, 41, 48, 60, 73, 102, 103, 111], although there exist some serious gaps between the theory and practical implementations. Thus, it is worth applying it to the control problem of OBMs. Furthermore, we have proposed a novel approach based on SMC by introducing a nonlinear sliding surface with variable-gain integral control, which will be a powerful tool to provide advantageous features with respect to transient-state control performance and control inputs when compared with the conventional SMC.

M. Toda, *Robust Motion Control of Oscillatory-Base Manipulators*,
Lecture Notes in Control and Information Sciences 463,
DOI 10.1007/978-3-319-21780-2_6

This chapter is organized as follows. Section 6.2 presents a brief introduction to the conventional SMC. In Sect. 6.3, we introduce our SMC-based approach and conduct stability analysis of the nonlinear sliding surface using the illustrative model as in the preceding chapters. Section 6.5 presents simulations to evaluate the proposed control system by comparing with a conventional SMC system, and the proposed $\mathcal{H}_\infty$ control-base approach in Chaps. 4 and 5, and finally, some concluding remarks are given.

6.2 Sliding-Mode Control

In this section, we briefly review the SMC framework. The SMC belongs to the wider class of variable structure systems (VSS), and its origin started in 1950s in USSR. Then, since Itkis [41] and Utkin [101] introduced the SMC theory to the world in the late 1970s, the SMC has gained a lot of attentions and become to be popular. Now, the SMC is well known to be one of the most powerful tools for robust control.

The SMC has unique features in the control law and control system structure, as the name of VSS suggests. The control law is a nonlinear and discontinuous function of the feedback variable, specifically a switching function. The system structure consists of two different structures called "*the reaching mode*" and "*the sliding mode*." Correspondingly to the structure, the design procedure also consists of two steps. First, the stable manifold in the state space of the system should be designed, on which manifold the trajectory necessarily converges to the origin asymptotically. Second, the control law to drive the system trajectory to the stable manifold and thereafter constrain it to the manifold should be chosen. Once the trajectory has been successfully confined to the manifold, the trajectory will automatically converge to the origin, like sliding on the manifold, which is the origin of the name, and the stable manifold is called the sliding surface.

In theory, the SMC is s considerably effective method for robust control, that is, to overcome problems due to disturbances and model uncertainties. However, there exist serious gaps between the theory and practical applications mainly due to the discontinuous nature of switching low which requires infinite-velocity switching. Hence, a direct implementation of the SMC in practice may often lead to unfavorable chattering phenomena and even worse instability of the system. Therefore, a lot of research efforts have been devoted to reduce or eliminate such gaps [25, 26], while other researchers have been trying to improve the SMC itself by considering a sliding surface, such as higher order [56], time-varying [11, 12], and nonlinear sliding surfaces [5, 32]. Our work also belongs to the latter works.

6.3 Sliding-Mode Control via Rotating Sliding Surface with Variable-Gain Integral Control

In this section, we address our proposed approach of sliding-mode control aiming at application to systems subject to strong disturbances such as an OBM. The approach is based on a nonlinear sliding surface proposed in [32] and additionally employs variable-gain integral control which will enhance the robustness of the control system against disturbances. The nonlinear sliding surface by [32] is originally motivated by the works of [11, 12] which proposed a sliding surface with the time-varying slope on the second-order phase plane, which is called a "rotating sliding surface." Reference [32] has arranged this idea by nonlinear tuning and inherited the term "rotating." Therefore, we also follow this stream and use the term "rotating sliding surface," when referring to the nonlinear sliding surface in our approach. Further, we refer to the rotating sliding surface with variable-gain integral control as RSSI, SMC via RSSI as SMC-RSSI, and conventional SMC via constant-gain integral control as SMC-I, respectively, for abbreviation. Those abbreviations will be used for both the respective control schemes and controllers using the schemes.

In this section, we present control system design of SMC-RSSI for the illustrative model of OBM mentioned in Chaps. 2 and 3.

6.3.1 Control System Design of SMC-RSSI

Here, we explain about control system design of SMC-RSSI using the problem setting in Chap. 2 and analyze a condition under which the convergence to the sliding surface is ensured, that is, "a sliding condition." Since the manipulator has two links, we design two SMC-RSSI systems for Links 1 and 2, respectively.

The dynamical model of the manipulator in (2.1) can be reformulated as

$$\dot{x}_1 = x_2$$

$$\dot{x}_2 = f_x + d_x + \gamma_x \tag{6.1}$$

$$\dot{z}_1 = z_2$$

$$\dot{z}_2 = f_z + d_z + \gamma_z \tag{6.2}$$

where x_1, z_1 denote the joint angles q_1 and q_2, respectively,

$$[f_x, f_z]^T = -M(\boldsymbol{q})^{-1}(C(\boldsymbol{q}, \dot{\boldsymbol{q}})\dot{\boldsymbol{q}} + D\dot{\boldsymbol{q}} + G(\boldsymbol{q}, \boldsymbol{q}_b))$$

are known functions of the states, $[d_x, d_z]^T = -M(\boldsymbol{q})^{-1}H(\boldsymbol{q}, \dot{\boldsymbol{q}}, \dot{\boldsymbol{q}}_b, \ddot{\boldsymbol{q}}_b)$ are unknown disturbances due to the base oscillation, and $[\gamma_x, \gamma_z]^T = M(\boldsymbol{q})^{-1}\boldsymbol{\tau}$ are the transformed control inputs.

The RSSIs are designed as

$$s_x = \dot{e}_x + a_x(e_x)e_x + b_x(e_x)\int_0^t e_x d\tau \tag{6.3}$$

$$s_z = \dot{e}_z + a_z(e_z)e_z + b_z(e_z)\int_0^t e_z d\tau \tag{6.4}$$

where a, b denote the proportional and integral control gains, respectively, $e_x = x_1 - r_1$, $e_z = z_1 - r_2$ are the tracking errors, r_1, r_2 are the joint angle references to be tracked as mentioned before, and t is the time variable. Note that variables with suffixes x, z denote ones associated with Links 1 and 2, respectively. a and b vary with the tracking error as follows.

$$a_x(e_x) = a_{xmin} + \frac{\delta a_x}{\cosh(er_{ax})} \tag{6.5}$$

$$a_z(e_z) = a_{zmin} + \frac{\delta a_z}{\cosh(er_{az})} \tag{6.6}$$

$$b_x(e_x) = b_{xmin} + \frac{\delta b_x}{\cosh(er_{bx})} \tag{6.7}$$

$$b_z(e_z) = b_{zmin} + \frac{\delta b_z}{\cosh(er_{bz})} \tag{6.8}$$

where

$$\delta a_x = a_{xmax} - a_{xmin} \tag{6.9}$$

$$\delta a_z = a_{zmax} - a_{zmin} \tag{6.10}$$

$$\delta b_x = b_{xmax} - b_{xmin} \tag{6.11}$$

$$\delta b_z = b_{zmax} - b_{zmin} \tag{6.12}$$

$$er_{ax} = \frac{e_x}{\sigma_{ax}} \tag{6.13}$$

$$er_{az} = \frac{e_z}{\sigma_{az}} \tag{6.14}$$

$$er_{bx} = \frac{e_x}{\sigma_{bx}} \tag{6.15}$$

$$er_{bz} = \frac{e_z}{\sigma_{bz}}, \tag{6.16}$$

the parameters with min and max are constants to determine the corresponding minimal and maximal gains, respectively, and σ_a, σ_b are constants to adjust the corresponding gain varying rates.

Remark 6.1 SMC-RSSI systems contain SMC-I ones in a natural manner such that the minimal and maximal gains are set to the same. Further, by setting δa and δb

to small enough, and/or σ to large enough, the SMC-RSSI system can be arbitrarily close to the corresponding SMC-I one.

To constrain the dynamics to the RSSIs $s_x = 0$ and $s_z = 0$, control inputs are designed as

$$\gamma_x = v_x - k_x \mathrm{sgn}(s_x) \tag{6.17}$$

$$\gamma_z = v_z - k_z \mathrm{sgn}(s_z) \tag{6.18}$$

$$\begin{aligned} v_x = &-f_x - \Big\{a_x \dot{e}_x + \frac{\delta a_x \sinh(er_{ax})}{\sigma_{ax} \cosh^2(er_{ax})} e_x \dot{e}_x \\ &+ b_x e_x + \dot{e}_x \frac{\delta b_x \sinh(er_{bx})}{\sigma_{bx} \cosh^2(er_{bx})} \int_0^t e_x d\tau \Big\} \end{aligned} \tag{6.19}$$

$$\begin{aligned} v_z = &-f_z - \Big\{a_z \dot{e}_z + \frac{\delta a_z \sinh(er_{az})}{\sigma_{az} \cosh^2(er_{az})} e_z \dot{e}_z \\ &+ b_z e_z + \dot{e}_z \frac{\delta b_z \sinh(er_{bz})}{\sigma_{bz} \cosh^2(er_{bz})} \int_0^t e_z d\tau \Big\} \end{aligned} \tag{6.20}$$

where k_x, k_z are positive constant gains such that $k_x > |d_x - \ddot{r}_1|$ and $k_z > |d_z - \ddot{r}_2|$.

Then, we analyze the sliding condition using the example of link 1. Note that the case of link 2 is exactly the same as that of link 1. Let

$$V_s(s_x) = \frac{1}{2} s_x^2 \tag{6.21}$$

be a Lyapunov function candidate for the sliding condition. Differentiating $V_s(s)$ with respect to t and substituting (6.1) and (6.17) yields

$$\begin{aligned} \dot{V}_s(s_x) &= s_x \dot{s}_x \\ &= s_x \Big\{\ddot{e}_x + a_x \dot{e}_x + \dot{a}_x e_x + b_x e_x + \dot{b}_x \int_0^t e_x d\tau \Big\} \\ &= s_x \Big\{\dot{x}_2 - \ddot{r}_1 + a_x \dot{e}_x + \dot{a}_x e_x + b_x e_x + \dot{b}_x \int_0^t e_x d\tau \Big\} \\ &= s_x \Big\{f_x + d_x + v_x - k_x \mathrm{sgn}(s_x) - \ddot{r}_1 + a_x \dot{e}_x \\ &\quad + \frac{\delta a_x \sinh(er_{ax})}{\sigma_{ax} \cosh^2(er_{ax})} e_x \dot{e}_x + b_x e_x \\ &\quad + \dot{e}_x \frac{\delta b_x \sinh(er_{bx})}{\sigma_{bx} \cosh^2(er_{bx})} \int_0^t e_x d\tau \Big\} \\ &= s_x \{d_x - \ddot{r}_1 - k_x \mathrm{sgn}(s_x)\} \\ &\le -(k_x - |d_x - \ddot{r}_1|)|s_x| \\ &< 0 \ (\text{for } s_x \neq 0) \end{aligned} \tag{6.22}$$

since $k_x > |d_x - \ddot{r}_1|$. Thus, the sliding condition is satisfied.

Next, we provide a perspective on how to choose the gain parameters of RSSI in (6.9)–(6.16). First, let us focus on the steady-state performance which is determined by the maximal gains. In this study, we assume that the references r_1 and r_2 are to be step, sinusoidal signals, and their linear sums. Thus, the integral gain b needs to be large so that the bandwidth is large enough for the given references. However, this should be generally done by taking into account of some constraints of practical implementation, i.e., the digital control sampling period, actuator limitations, and sensor noises. Then, the proportional gain a should be chosen under consideration of an appropriate balance with the b. Meanwhile, the transient performance can be modified by the minimal gains and gain varying rates. By properly choosing those parameters, the transient performance can be improved in comparison with the SMC-I controller, which is demonstrated later by simulations.

On the other hand, k_x and k_z in (6.17) and (6.18) are the most important parameters in order to constrain the dynamics to the sliding surface as proven above. Theoretically and without constrains of practical implementation, the lager k_x and k_z lead to the stronger robustness against disturbances and model uncertainties which satisfy the matching condition. Their appropriate values depend on the problems and situations. As demonstrated by simulations, appropriate values of k_x and k_z in the nominal-case problem and in the robust control problem are largely different, because in the robust control problem d_x and d_z contain not only the disturbance due to the base oscillation but also the error of canceling the nonlinearity f_x and f_z.

6.3.2 Control System Design Example

Table 6.1 shows controller parameters of an example of SMC-RSSI which is utilized to demonstrate stability analysis of RSSI in the next section. Using this SMC-RSSI, simulations of attitude control in local coordinates in the case of single-frequency oscillation were conducted. In Fig. 6.1, the upper graph shows the result with the sampling period of 0.01 s while the lower one does that with the sampling period of 0.001 s. The simulation conditions are the same as those for the $\mathcal{H}_\infty$ controllers in Chap. 5 except the sampling time. First, the SMC-RSSI reveals successful performances in the both sampling period cases. On the other hand, we see that the controller is largely influenced by the sampling period, which is the characteristic feature of SMC. This fact is natural because in theory the SMC scheme assumes an infinite velocity switching control inputs.

Table 6.1 Controller parameters of an example of SMC-RSSI

Link	a_{min}	a_{max}	σ_a	b_{min}	b_{max}	σ_b	k
1	5	40	0.7	80	350	0.7	6
2	5	80	0.7	100	700	0.7	4

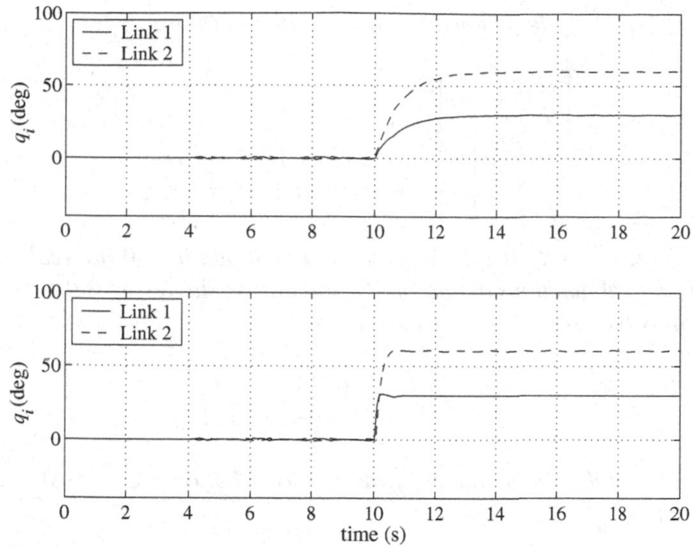

Fig. 6.1 Simulation results of attitude control in local coordinates with the SMC-RSSI example for the single-frequency oscillation; *top* with the sampling period of 0.01 s, *bottom* with the sampling period of 0.001 s

6.4 Stability Analysis of RSSI

The time-varying sliding surface in [11, 12] and the nonlinear sliding surface in [32] are scalar systems and hence their stabilities are guaranteed as long as their slopes are negative. However, the RSSI is a nonlinear second-order system and the problem of its stability is nontrivial. Therefore, we need stability analysis of RSSI to confirm that the RSSI is a stable manifold.

In this monograph, we analyze two cases of global stability and semi-global stability, respectively. To analyze stability of RSSI, we prepare a lemma which provides a sufficient condition to construct a Lyapunov function. If this condition is globally satisfied on the RSSI, then the origin of the RSSI is globally asymptotically stable.

Otherwise its semi-global stability can be proved by conducting the following procedures. First, we differentiate regions on the RSSI where the condition in the lemma is satisfied and not satisfied, respectively. Second, by utilizing the Lyapunov function and a numerical solution, we show that there exists an invariant compact set containing the origin as a unique equilibrium for trajectories on the RSSI. Finally, by showing that any closed orbits, that is, periodical solutions do not exist in the invariant compact set, it is proved that the RSSI is asymptotically stable with respect to the origin within the set. Respective theorems regarding the global stability and semi-global stability of RSSI are given.

Here, we prepare the following lemma which provides a machinery to construct a Lyapunov function for a time-varying and/or nonlinear second-order system.

Lemma 6.1 *Consider the following second-order system,*

$$\begin{aligned}\begin{bmatrix} \dot{ie} \\ \dot{e} \end{bmatrix} &= A_e \begin{bmatrix} ie \\ e \end{bmatrix} \\ &= \begin{bmatrix} 0 & 1 \\ -(b+\delta b) & -(a+\delta a) \end{bmatrix} \begin{bmatrix} ie \\ e \end{bmatrix} \end{aligned} \tag{6.23}$$

where $\mathbf{e}' = [ie, e]^T \in D$ is the state vector, $a > 0$ and $b > 0$ are real constants, and δa, δb are real functions of time and/or the state variables. D is the real domain which contains the origin. Suppose the matrix

$$A_{e0} = \begin{bmatrix} 0 & 1 \\ -b & -a \end{bmatrix} \tag{6.24}$$

has a pair of complex conjugate eigenvalues $-\alpha \pm i\beta$, $\alpha > 0$, $\beta > 0 \in \mathbb{R}$. Then, using the matrix

$$S_e = \begin{bmatrix} 1 & 1 \\ -\alpha + \beta & -\alpha - \beta \end{bmatrix}, \tag{6.25}$$

$$V(\mathbf{e}') = \mathbf{e}'^T (S_e^{-1})^T S_e^{-1} \mathbf{e}' \tag{6.26}$$

can be a Lyapunov function for the system in (6.23) with $\delta a = 0$ and $\delta b = 0$. Further, let $\delta p = \delta b - \alpha\delta a$, and if

$$\delta a > \frac{1}{4\alpha\beta^2}\delta p^2 - \alpha \tag{6.27}$$

over D, then $V(\mathbf{e}')$ is also a Lyapunov function for the system in (6.23).

Proof Since $|S_e| = -2\beta \neq 0$, S_e is nonsingular and hence can be a similarity transformation to the system in (6.23). Then, A_e is transformed into

$$P_e = S_e^{-1} A_e S_e.$$

For A_{e0} we use P_{e0} accordingly.

Further, $V(\mathbf{e}')$ is positive definite. Thus, differentiating $V(\mathbf{e}')$ with respect to time yields

$$\dot{V}(\mathbf{e}') = \mathbf{e}'^T (S_e^{-1})^T (P_e^T + P_e)(S_e^{-1})\mathbf{e}'.$$

First, we consider the case where $\delta a = 0$ and $\delta b = 0$. Then,

$$P_{e0}^T + P_{e0} = \begin{bmatrix} -2\alpha & 0 \\ 0 & -2\alpha \end{bmatrix} \prec 0,$$

which proves that $\dot{V}(\boldsymbol{e}')$ is negative definite, together with the nonsingularity of S_e. Thus, $V(\boldsymbol{e}')$ is a Lyapunov function.

Next, we investigate the case where $\delta a \neq 0$ and/or $\delta b \neq 0$. In this case, by some algebraic manipulations with

$$a = 2\alpha,$$
$$b = \alpha^2 + \beta^2,$$

we obtain

$$P_e^T + P_e = -\frac{1}{\beta}\begin{bmatrix} 2\alpha\beta + \delta a(-\alpha+\beta) + \delta b & -\delta a\beta \\ -\delta a\beta & 2\alpha\beta + \delta a(\alpha+\beta) - \delta b \end{bmatrix}.$$

Being based on Sylvester's criterion, this matrix is negative definite, if and only if the following inequalities hold

$$\delta a > -\frac{1}{\beta}\delta p - 2\alpha,$$
$$\delta a > -\frac{1}{4\alpha\beta^2}\delta p^2 - \alpha,$$

where $\delta p = \delta b - \alpha\delta a$. However, since

$$\{-\frac{1}{4\alpha\beta^2}\delta p^2 - \alpha\} - \{-\frac{1}{\beta}\delta p - 2\alpha\} = \frac{1}{4\alpha\beta}(\delta p + 2\alpha\beta)^2 \geq 0$$

those conditions can be integrated into the inequality (6.27). Therefore, if (6.27) holds, then $V(\boldsymbol{e}')$ is a Lyapunov function for the system in (6.23), and the proof is completed.

Note that e and ie correspond to the tracking error and its integral on the RSSI, and that the complex eigenvalue condition is the case for the RSSI to be analyzed.

Now, we demonstrate two cases of stability analysis by deploying the example of SMC-RSSI in Table 6.1. First, we derive the following theorem regarding the global stability of RSSI for the case of Link 1 with the SMC-RSSI.

Theorem 6.1 (Global stability) *Consider the RSSI represented by the following second-order system.*

$$\begin{bmatrix} \dot{ie} \\ \dot{e} \end{bmatrix} = \begin{bmatrix} 0 & 1 \\ -b(e) & -a(e) \end{bmatrix}\begin{bmatrix} ie \\ e \end{bmatrix} \tag{6.28}$$

where the domain D of the state vector $\boldsymbol{e}' = [ie, e]^T$ *is* $\mathbb{R} \times [-\pi, \pi]$, $a(e) = a_x(e_x)$ *in (6.5),* $b(e) = b_x(e_x)$ *in (6.7), and the associated parameters are defined as those for Link 1 with the SMC-RSSI in Table 6.1. Then, the RSSI is globally asymptotically stable with respect to the origin.*

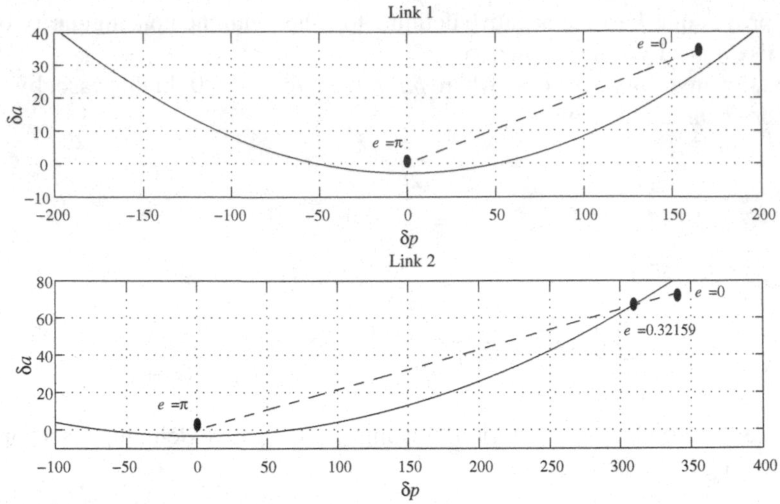

Fig. 6.2 Results of checking the condition for a Lyapunov function in (6.27) for the SMC-RSSI in Table 6.1; *solid line* boundary for the condition in (6.27), *dashed line* parameter trajectory with respect to e, *top* Link 1, *bottom* Link 2

Proof Since it can be straightforwardly confirmed that the right-hand side function in (6.28) is globally Lipschitz in $\boldsymbol{e}'$, the existence and uniqueness of the solution to every initial point $\boldsymbol{e}'(0) \in D$ is ensured. Further, due to the nonsingularity of the right-hand side matrix, the origin is the unique equilibrium.

Then, we construct a Lyapunov function candidate with $e = \pi$ and check the sufficient condition in (6.27) with Lemma 6.1. The result is shown in the top graph in Fig. 6.2, where the dashed line represents $(\delta p, \delta a)$ over $e \in [0, \pi]$ and the solid line does the boundary by (6.27). Note that $a(\cdot)$ and $b(\cdot)$ are even functions with respect to e. Thus, it is seen that the condition in (6.27) is globally satisfied on the domain D. Therefore, the proof is completed.

Next, we present a semi-global stability case with the example of Link 2 as in the following theorem.

Theorem 6.2 (Semi-global stability) *As in Theorem 6.1, consider the RSSI represented by (6.28) over the same domain and the corresponding parameters are defined as those for Link 2 with the SMC-RSSI in Table 6.1. Then, the RSSI is semi-globally asymptotically stable with respect to the origin.*

Proof The existence and uniqueness of the solution and the property of the origin on the RSSI are the same as in Theorem 6.1. We build a Lyapunov function candidate with $e = \pi$ and check the condition by (6.27). The bottom graph in Fig. 6.2 shows the result. It is seen that the condition is satisfied for $0.32159 \leq |e| \leq \pi$, however is not for $0 \leq |e| < 0.32159$.

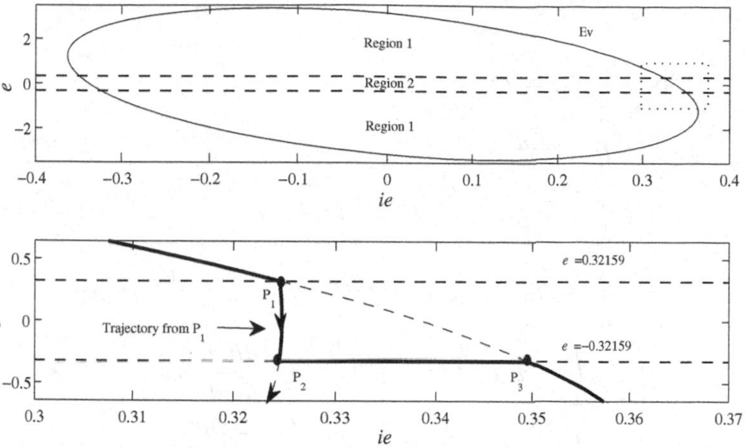

Fig. 6.3 Invariant compact set on the phase plane; *top* on the normal scale, *bottom* zoomed-in area encircled by the *dotted rectangle* in the *top figure*

Then, we show that there exists an invariant compact set of the RSSI. To this end, we define regions on the domain D as $\{ie, e|ie \in \mathbb{R}, 0.32159 \leq |e| \leq \pi\}$ (Region 1) where the condition in (6.27) is satisfied and $\{ie, e|ie \in \mathbb{R}, 0 \leq |e| \leq 0.32159\}$ (Region 2) where the condition is not satisfied.

Next, we define $V(\boldsymbol{e}') = const$ such that the corresponding ellipsoid Ev on the ie-e phase plane passes through $(0, \pi)$, which is depicted in Fig. 6.3 (see the top one). Focus on the area encircled by the dotted rectangle which is magnified on the bottom figure in Fig. 6.3. By calculating the solution with the initial point P_1, we can obtain the trajectory from P_1 which passes through Region 2 and goes into Region 1 inside Ev. On the other hand, the trajectory from every point on the line segment $\mathrm{P}_2\mathrm{P}_3$ also goes into Region 1 inside Ev according to the vector field. Considering those facts and the uniqueness of the solution together with the symmetry of the vector field by (6.28) with respect to the origin, therefore, it is concluded that the compact set with the boundary consisting of the parts of Ev in Region 1, the trajectory from P_1 to P_2, the line segment $\mathrm{P}_2\mathrm{P}_3$ and their symmetric counterparts is an invariant set. That is, the trajectory from every point in the set is constrained in the set.

Hence, by applying Poincare-Bendixson theorem, if there exists no periodic solution in this compact set, it is ensured that the compact set is asymptotically stable with the origin. We show which by contradiction.

Suppose that there exists a periodic solution $\boldsymbol{e}'(t) = \boldsymbol{e}'(t + T)$ in the compact set, and define the energy function

$$E(\boldsymbol{e}') = \frac{1}{2}b(e)ie^2 + \frac{1}{2}e^2. \tag{6.29}$$

Then,

$$
\begin{aligned}
0 &= E(e'(t+T)) - E(e'(t)) \\
&= \int_t^{t+T} \dot{E} d\tau \\
&= \int_t^{t+T} \{b(e)ie\dot{i}e + e\dot{e} + \frac{1}{2}\dot{b}ie^2\} d\tau \\
&= \int_t^{t+T} \{b(e)iee + e(-b(e)ie - a(e)e) + \frac{1}{2}\dot{b}ie^2\} d\tau \\
&= -\int_t^{t+T} a(e)e^2 d\tau + \frac{1}{2}\int_t^{t+T} ie^2 \frac{db}{de}\frac{de}{d\tau} d\tau \\
&= -\int_t^{t+T} a(e)e^2 d\tau + \frac{1}{2}\int_{e(t)}^{e(t+T)} ie^2 \frac{db}{de} de \\
&= -\int_t^{t+T} a(e)e^2 d\tau \\
&< 0
\end{aligned}
\tag{6.30}
$$

which is contradiction. Thus, the proof is completed.

Note that this asymptotically stable region is sufficiently large for a practical application.

6.5 Simulations

In order to evaluate the proposed SMC-RSSI systems, control simulations were performed emphasizing comparison with the proposed $\mathcal{H}_\infty$ control systems. The demonstration cases are the same as in the case of $\mathcal{H}$ control in Chap. 5. First, for the nominal case, i.e., the payload width is 5 mm, all the combinations of the three types of control problems and three types of base oscillation will be presented. Then, robust control demonstrations will be addressed. In all the demonstrations, the methods and conditions are almost the same as those presented in Chap. 5 except the sampling period and the sensor resolution. As illustrated in Fig. 6.1, SMC systems are largely influenced by the sampling period. Even in the nominal case, in order that the SMC-RSSI performs satisfactorily for those various demonstration cases, the sampling period of 0.001 s is required. Moreover, the control system is also affected by the sensor resolution. Two sensor resolutions of 0.18 and 0.0018 deg are compared for the simulations, to investigate such properties of the system and show that how effective the system can be under desirable conditions of sensor resolution. For the simulation cases, the control parameters were tuned through a trial and error manner, which are shown in Table 6.2.

Table 6.2 Controller parameters of SMC-RSSIs for simulation cases

Case	Link	a_{min}	a_{max}	σ_a	b_{min}	b_{max}	σ_b	k
Nominal	1	25	25	–	1	4000	0.3	6
	2	30	30	–	1	1500	0.7	4
Robust	1	25	25	–	1	4000	0.3	12
	2	30	30	–	1	1500	0.7	45

6.5.1 Simulations in the Nominal Case

Table 6.3 compares the simulation results of the SMC-RSSI with 0.18 deg sensor resolution, the SMC-RSSI with 0.0018 deg one, the $\mathcal{H}_\infty$ controllers with 0.18 deg one, and the $\mathcal{H}_\infty$ controllers with 0.0018 deg one, with respect to RMSE $\bar{e}$. Note that those controllers were performed with the sampling period of 0.001 s except the $\mathcal{H}_\infty$ controllers with 0.18 deg sensor resolution (with that of 0.01 s).

First, let us compare the results of the SMC-RSSI and $\mathcal{H}_\infty$ controllers in the case of sensor resolution of 0.18 deg. Except the position control cases, the $\mathcal{H}_\infty$ controllers performed slightly better the SMC-RSSI. Further, it should be noted that this SMC-RSSI requires five times larger toque limits than those for the other controllers.

Then, we review the results in the case of finer sensor resolution of 0.0018 deg. Interestingly enough, the results of the SMC-RSSI has been considerably improved with the finer sensor resolution. Similarly, those of the $\mathcal{H}_\infty$ controllers has been

Table 6.3 Results of RMSE $\bar{e}$ in the respective control simulations

Single	SMC-RSSI (0.18 deg)	SMC-RSSI (0.0018 deg)	K_{21} (0.18 deg)	K_{21} (0.0018 deg)
Attitude-l (deg)	0.2088	0.1643	0.1047	0.0557
Attitude-g (deg)	0.2131	0.1500	0.2152	0.2034
Position (mm)	0.2336	0.0559	0.3972	0.3846
Double	SMC-RSSI (0.18 deg)	SMC-RSSI (0.0018 deg)	K_{22} (0.18 deg)	K_{22} (0.0018 deg)
Attitude-l (deg)	0.1690	0.0318	0.1072	0.0630
Attitude-g (deg)	0.1775	0.0346	0.1361	0.0721
Position (mm)	0.2073	0.0137	0.2276	0.1410
Bretschneider	SMC-RSSI (0.18 deg)	SMC-RSSI (0.0018 deg)	K_{23} (0.18 deg)	K_{23} (0.0018 deg)
Attitude-l (deg)	0.1773	0.0046	0.1085	0.0163
Attitude-g (deg)	0.1988	0.0056	0.1486	0.0401
Position (mm)	0.2712	0.0133	0.2358	0.1283

Attitude-l attitude control in local coordinates, *attitude-g* attitude control in global coordinates, *position* position control. Note that the SMC-RSSI with 0.18 deg resolution requires five times larger torque amplitude limits

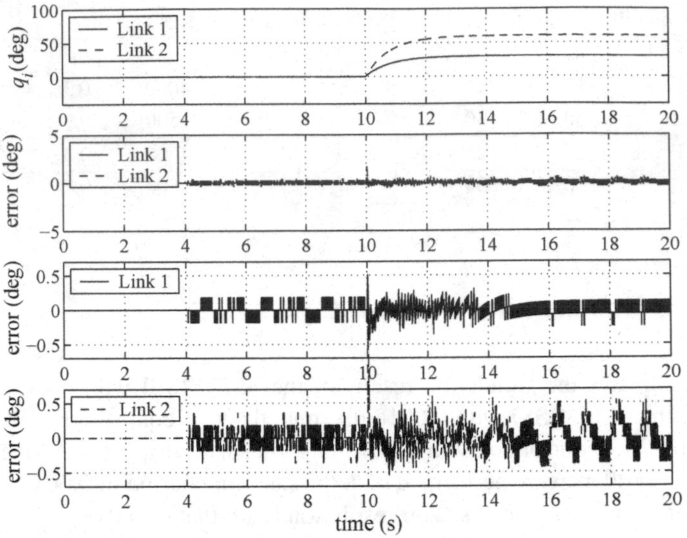

Fig. 6.4 Simulation results of attitude control in local coordinates with the SMC-RSSI (0.18 deg) for the double-frequency oscillation; *top* the attitudes of the links in the local coordinates, second from the *top*: the tracking errors on a normal scale, third from the *top*: the tracking error of Link 1 on a fine scale, *bottom* the tracking error of Link 2 on a fine scale

improved, but the degree of improvement is less than that of the SMC-RSSI. In other words, the SMC-RSSI tends to be influenced by sensor resolution and noise in terms of control performance and required control inputs, being compared to the $\mathcal{H}_\infty$ controllers. On the other hand we should note that for various types of base oscillation the single SMC-RSSI can work well, whereas the corresponding weighting functions to the assumed base oscillation should be prepared to design an $\mathcal{H}_\infty$ controller.

Next, we present the results in time-series graphs by choosing examples. The double-frequency demonstrations have been selected to display the time-series profiles of control performance. For each control problem, the control performances of the SMC-RSSIs with the sensor resolutions of 0.18 and 0.0018 deg are compared. Figures 6.4 and 6.5 present the simulation results of attitude control in local coordinates with the respective SMC-RSSIs. Then, Figs. 6.6 and 6.7 show the results of attitude control in global coordinates, and Figs. 6.8 and 6.9 show those of position control. As seen from the results, both the SMC-RSSIs suppress the disturbances and achieve successful control performance. In particular, by comparing the control performances with those of the $\mathcal{H}_\infty$ controllers, not only the steady-state performances but also the transient performances at just after 10 s are considerably successful, which is exactly the benefit of the RSSI as demonstrated later. Furthermore, it is seen that such a finer sensor resolution can dramatically improve the control performance of the SMC-RSSI.

By changing the view point, we examine profiles of control inputs of the SMC-RSSIs which represent the characteristic feature of SMC schemes. The results of

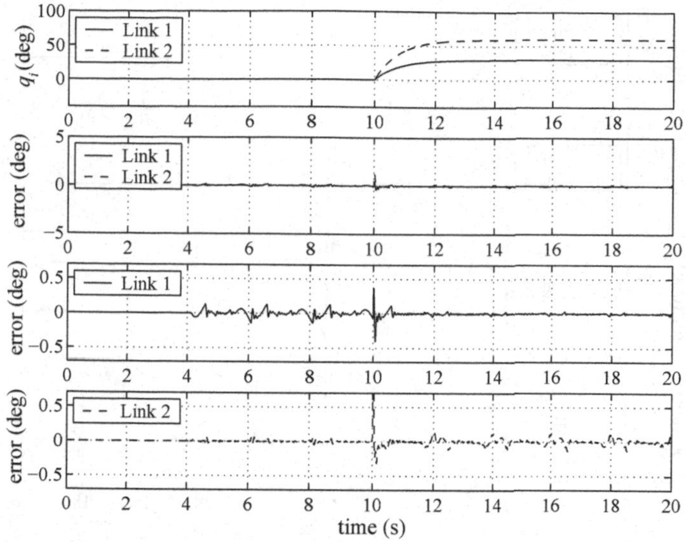

Fig. 6.5 Simulation results of attitude control in local coordinates with the SMC-RSSI (0.0018 deg) for the double-frequency oscillation; *top* the attitudes of the links in the local coordinates, second from the *top*: the tracking errors on a normal scale, third from the *top*: the tracking error of Link 1 on a fine scale, *bottom* the tracking error of Link 2 on a fine scale

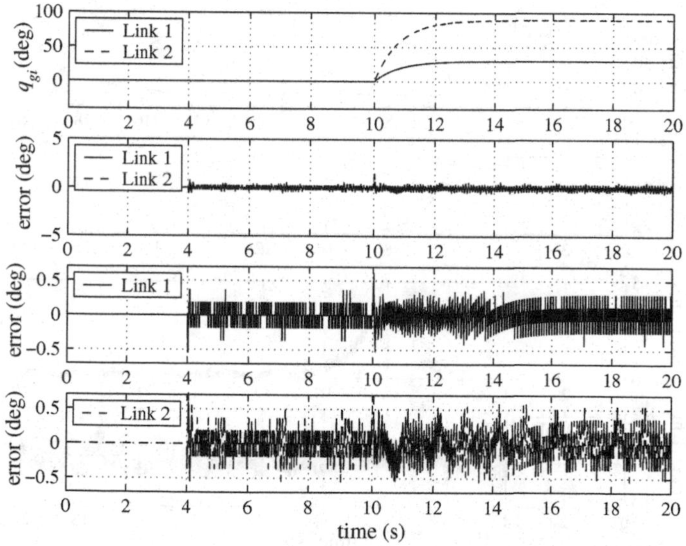

Fig. 6.6 Simulation results of attitude control in global coordinates with the SMC-RSSI (0.18 deg) for the double-frequency oscillation; *top* the attitudes of the links in the local coordinates, second from the *top*: the tracking errors on a normal scale, third from the *top*: the tracking error of Link 1 on a fine scale, *bottom* the tracking error of Link 2 on a fine scale

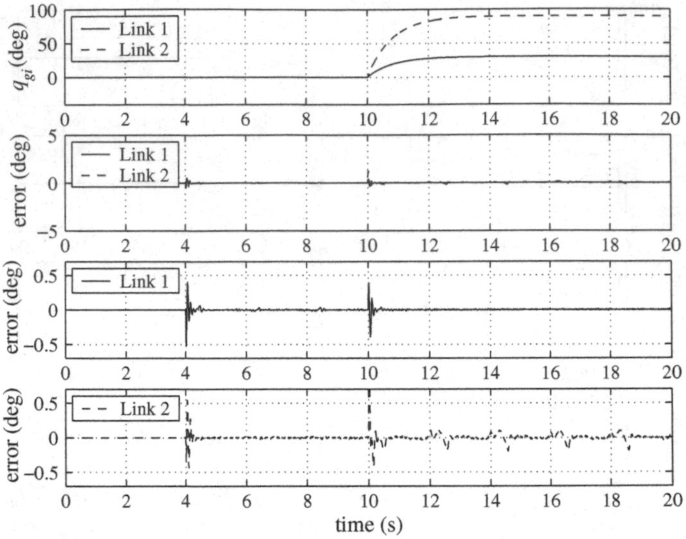

Fig. 6.7 Simulation results of attitude control in global coordinates with the SMC-RSSI (0.0018 deg) for the double-frequency oscillation; *top* the attitudes of the links in the local coordinates, second from the *top*: the tracking errors on a normal scale, third from the *top*: the tracking error of Link 1 on a fine scale, *bottom* the tracking error of Link 2 on a fine scale

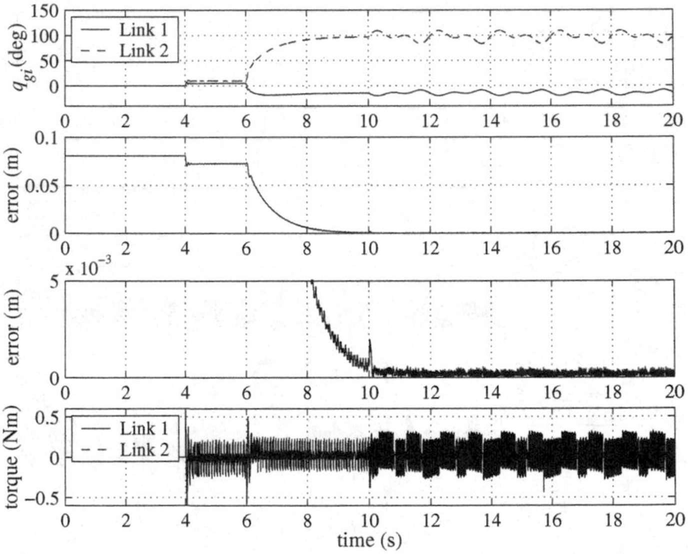

Fig. 6.8 Position control simulation results with the SMC-RSSI (0.18 deg) for the double-frequency oscillation; *top* the attitudes of the links in the global coordinates, second from the *top*: the distance errors from the final goal on a normal scale, third from the *top*: the distance errors from the final goal on a fine scale, *bottom* the control torques

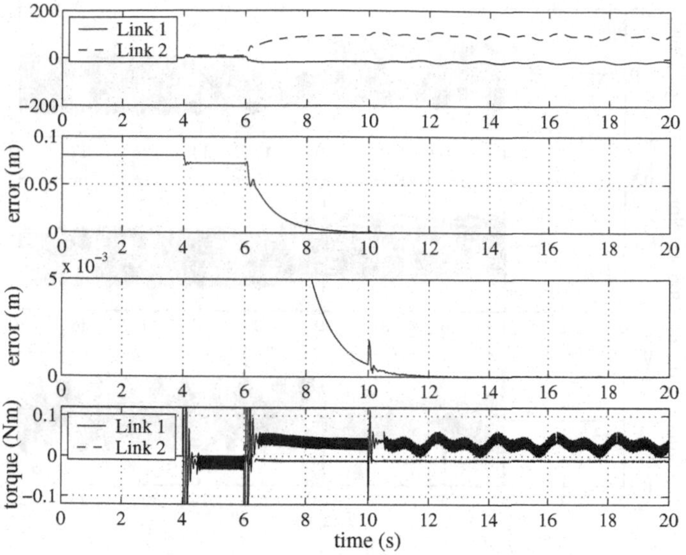

Fig. 6.9 Position control simulation results with the SMC-RSSI (0.0018 deg) for the double-frequency oscillation; *top* the attitudes of the links in the global coordinates, second from the *top*: the distance errors from the final goal on a normal scale, third from the *top*: the distance errors from the final goal on a fine scale, *bottom* the control torques

control toque inputs in the simulations of the respective control problems, links, and SMC-RSSIs are shown in Figs. 6.10, 6.11, 6.12, 6.13, respectively. First, those profiles exhibit features like thick belts when compared to those of the $\mathcal{H}_\infty$ controller which seems to be thin strings (see Figs. 5.14 and 5.15 again), which is exactly the characteristic feature of switching control by SMC. Then, comparing those of the SMC-RSSIs with 0.18 and 0.0018 deg, Figs. 6.10 and 6.11 (case of 0.18 deg) reveal five times larger torques than those in Figs. 6.12 and 6.13 (case of 0.0018 deg). In particular, in the case of 0.18 deg, the required switching amplitude might be tough for practical implementation to actuators. We should note that this characteristic feature of control input give rise to the serious gap between the theory and practical application of the SMC scheme, such as chartering phenomena, unexpected control errors.

Here, we address effectiveness of the RSSI being compared to the conventional constant-gain sliding surface. To this end, we performed step-tracking attitude control in local coordinates for the double-frequency oscillation for the SMC-RSSI and a SMC-I without any low-pass filter for the step signal, so that the effectiveness of the RSSI becomes apparent. The SMC-I employed the maximal gains of the SMC-RSSI (in the nominal case in Table 6.2) as its constant gains. First, Fig. 6.14 compares their results in the case of unbounded control torques. As seen from the figure, the both results exhibit satisfactory steady-state performances similarly while the SMC-I system suffers from large overshoot during the transient state. Figure 6.15 display

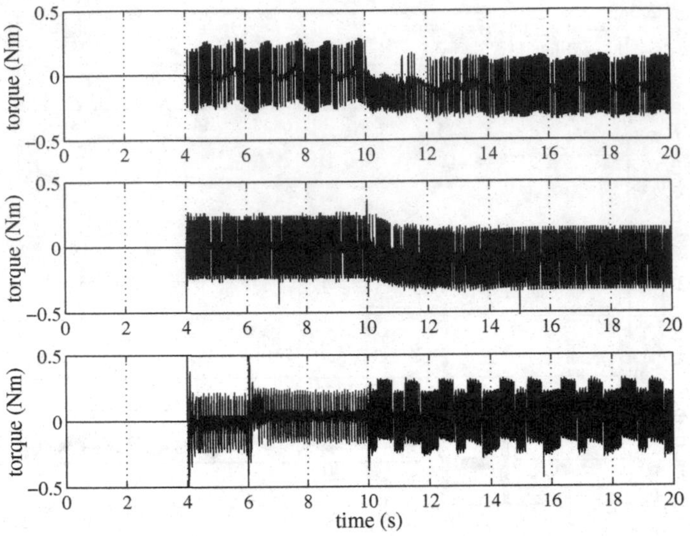

Fig. 6.10 Control torques of the joint 1 with the SMC-RSSI (0.18 deg) for the respective problems in the case of double-frequency oscillation; *top* attitude control in local coordinates, *middle* attitude control in global coordinates, *bottom* position control in global coordinates

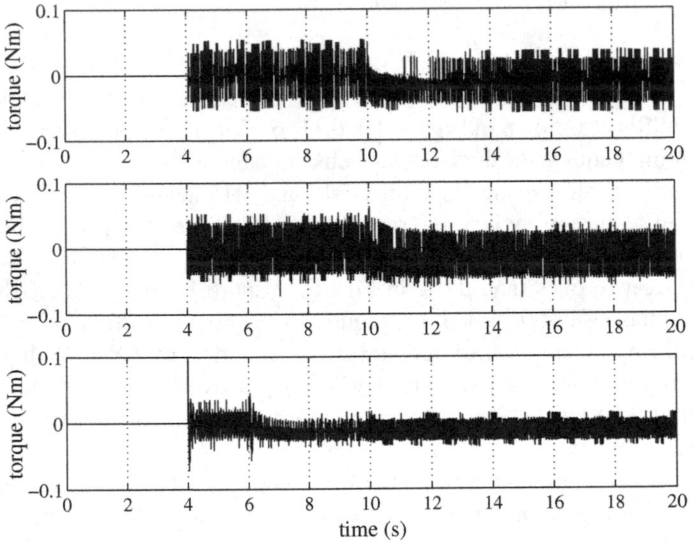

Fig. 6.11 Control torques of the joint 2 with the SMC-RSSI (0.18 deg) for the respective problems in the case of double-frequency oscillation; *top* attitude control in local coordinates, *middle* attitude control in global coordinates, *bottom* position control in global coordinates

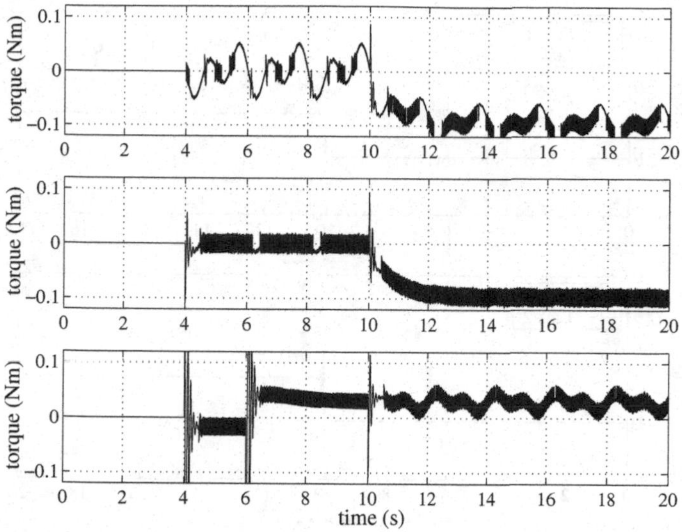

Fig. 6.12 Control torques of the joint 1 with the SMC-RSSI (0.0018 deg) for the respective problems in the case of double-frequency oscillation; *top* attitude control in local coordinates, *middle* attitude control in global coordinates, *bottom* position control in global coordinates

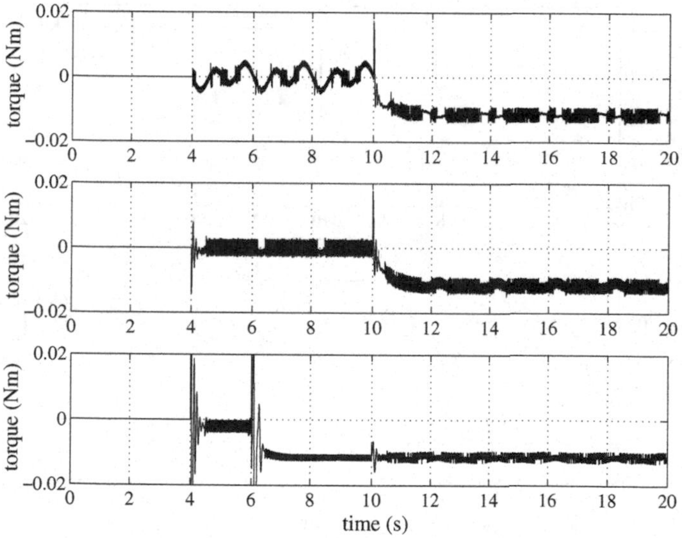

Fig. 6.13 Control torques of the joint 2 with the SMC-RSSI (0.0018 deg) for the respective problems in the case of double-frequency oscillation; *top* attitude control in local coordinates, *middle* attitude control in global coordinates, *bottom* position control in global coordinates

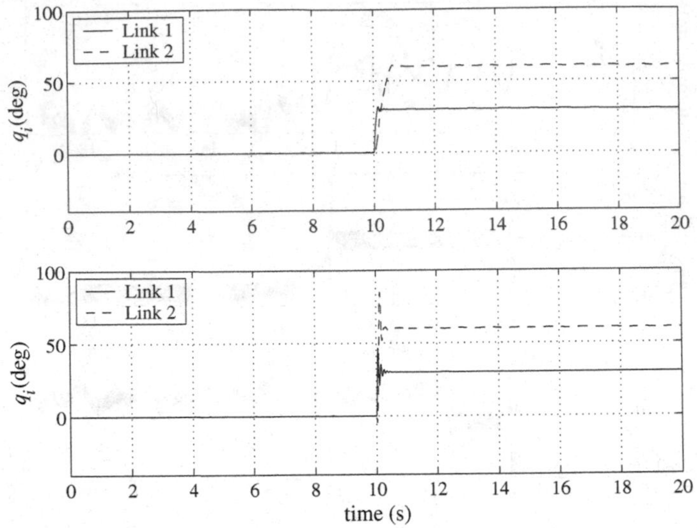

Fig. 6.14 Comparison of simulation results with the SMC-RSSI and SMC-I in the case of unbounded control torque (attitude control in local coordinates for the double-frequency oscillation); *top* SMC-RSSI, *bottom* SMC-I

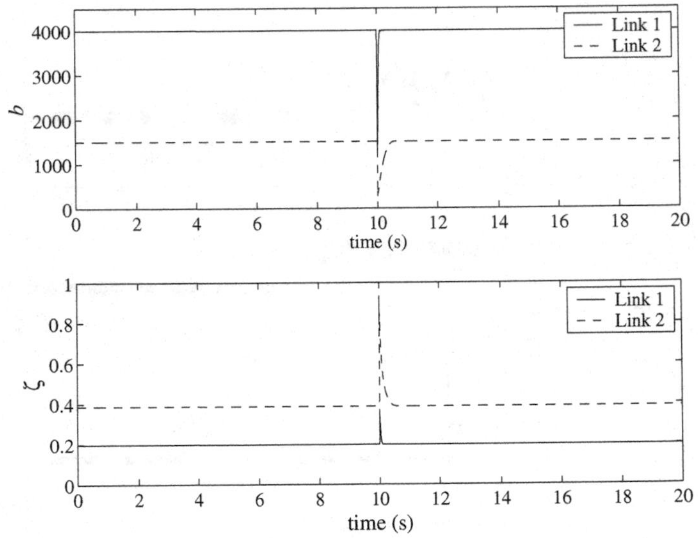

Fig. 6.15 Variation of b_x, b_z, and the damping ratios of simulation results with the SMC-RSSI and SMC-I in the case of unbounded control torque (attitude control in local coordinates for the double-frequency oscillation); *top* SMC-RSSI, *bottom* SMC-I

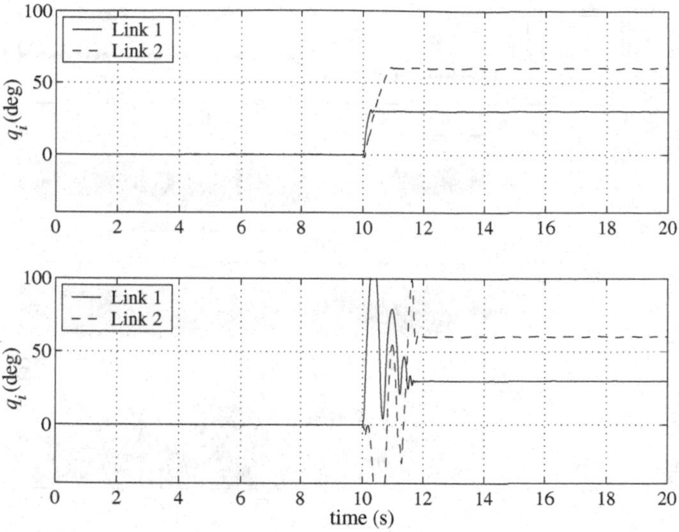

Fig. 6.16 Comparison of simulation results with the SMC-RSSI and SMC-I in the case of bounded control torque (attitude control in local coordinates for the double-frequency oscillation); *top* SMC-RSSI, *bottom* SMC-I

how the SMC-RSSI varies the proportional gains b_x and b_z and correspondingly the damping ratios in the control simulation. The gains decrease steeply and the damping ratios increase accordingly, which plays a key role in suppressing the overshoot and control inputs.

The point is that when the tracking error is large, the controller can accelerate the object enough even with a small proportional gain and even tries to suppress too large velocity and undesirable large surge control input, then as the error is getting small the controller tries to keep a moderate velocity by increasing the proportional gain accordingly within the range which will not lead to overshooting. Which enables the controller to achieve such a rapid response in control performance without overshooting and too large control inputs.

Next, we examine the results of the same simulation in the case of bounded control torques similarly to that of the $\mathcal{H}_\infty$ controllers and most of the SMC-RSSIs mentioned before. Figure 6.16 presents the results, where the SMC-RSSI reveals a similar successful performance to that in the case of unbounded control torque, but with a slightly slower response, whereas the SMC-I does unacceptably bad transient-state performance. Therefore, all the above arguments have proved that the RSSI can provide effectiveness in improving transient-state control performance and suppressing control inputs.

Finally, in this section, we demonstrate how the SMC-RSSI is largely influenced by the sensor resolution and sampling period. Figure 6.17 shows the simulation result of attitude control in global coordinates for the double-frequency oscillation with the SMC-RSSI with the sensor resolution of 0.18 deg and the sampling period of 0.01s,

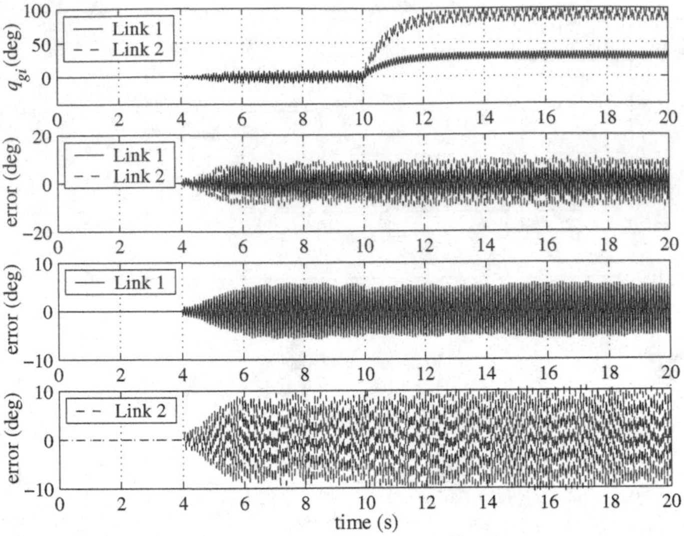

Fig. 6.17 Simulation results of attitude control in global coordinates with the SMC-RSSI (0.18 deg) with the sampling period of 0.01 s for the double-frequency oscillation; *top* the attitudes of the links in the local coordinates, second from the *top*: the tracking errors on a normal scale, third from the *top*: the tracking error of Link 1 on a fine scale, *bottom* the tracking error of Link 2 on a fine scale

where it is seen that such a condition deteriorate the control performance of the SMC-RSSI seriously. Which is also one of features representing the gap between the theory and practical application of the SMC scheme.

6.5.2 Simulations in the Robust Control Case

In this section, we present simulation results in the robust control case, where all the combinations of the three types control problems and three types of base oscillations with the payload variation with the widths, 0, 1, 2, ..., 9, 10 mm were conducted. Not like in the nominal-case demonstrations, we did not use the SMC-RSSI with the sensor resolution of 0.18 because the controller cannot achieve acceptable roust control performance. Moreover, since the SMC-RSSI (0.0018 deg, $k_x = 6$, $k_z = 4$) cannot satisfactory robust control performance compared to those of the $\mathcal{H}_\infty$ controllers, we prepared a new SMC-RSSI of which robustness has been enhanced by increasing k_x to be 12 and k_z to be 45. As shown in Table 6.2, the other parameters are the same.

As in the case of the $\mathcal{H}_\infty$ controllers, the simulation results are expressed being based on the RMSE $\bar{e}$'s versus the payload widths in the respective control problems. Figures 6.18 and 6.19 show the results of the robust control simulations with the

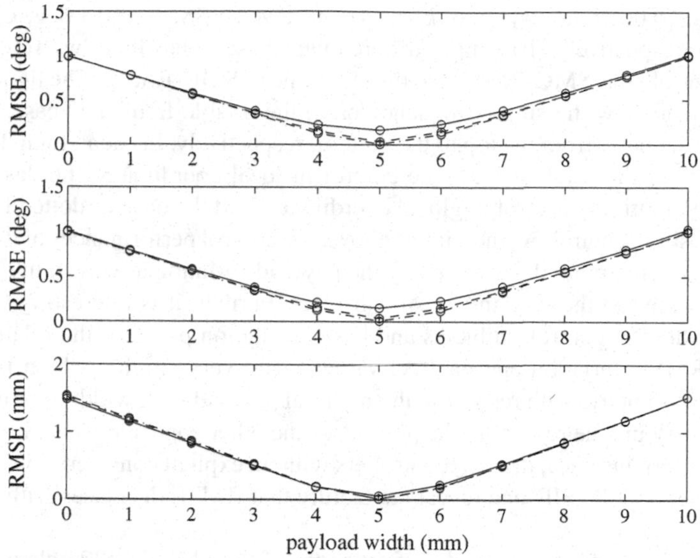

Fig. 6.18 RMSE $\bar{e}$ versus payload width of the simulation results of robust control with the SMC-RSSI (0.0018 deg, $k_x = 6, k_z = 4$); *top* attitude control in local coordinates, *middle* attitude control in global coordinates, *bottom* position control in global coordinates; *solid* single frequency, *dashed* double frequency, *dashed-dotted* Bretschneider

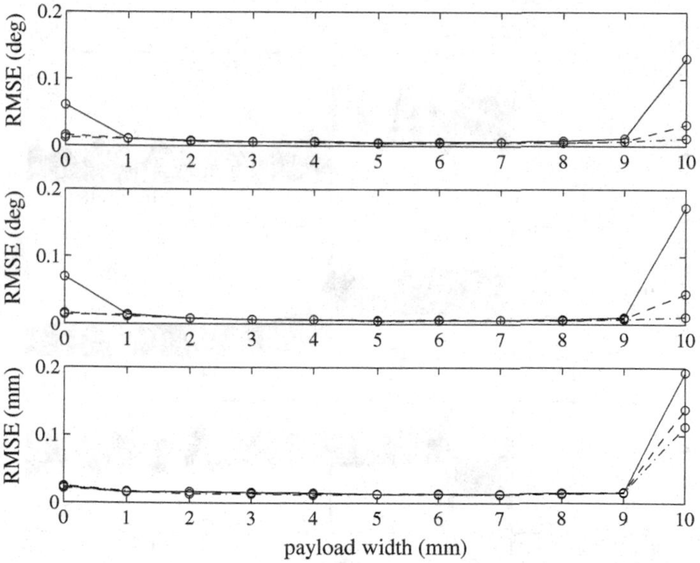

Fig. 6.19 RMSE $\bar{e}$ versus payload width of the simulation results of robust control with the SMC-RSSI (0.0018 deg, $k_x = 12, k_z = 45$); *top* attitude control in local coordinates, *middle* attitude control in global coordinates, *bottom* position control in global coordinates; *solid* single frequency, *dashed* double frequency, *dashed-dotted* Bretschneider

SMC-RSSI (0.0018 deg, $k_x = 6$, $k_z = 4$) and SMC-RSSI (0.0018 deg, $k_x = 12$, $k_z = 45$), respectively. To simply differentiate those controllers, we refer to the former controller as SMC-RSSI1 and the latter one as SMC-RSSI2. The figures have graphs which show the single-frequency case, the double-frequency case, and the Bretschneider case, from the top to the bottom, respectively. In each graph, the solid line represents the results of attitude control in local coordinates, the dashed line does those of attitude control in global coordinates, and the dashed-dotted line does those of position control. As mentioned above, the control performances of the SMC-RSSI1 are considerably deteriorated by the payload variation as shown in Fig. 6.18, being compared to those of the $\mathcal{H}_\infty$ controllers. Further, it is interesting to notice that, regardless of control problems and base-oscillation patterns, the relationships between RMSE variation and payload variation are very similar, which is almost linear and symmetric with respect to the nominal payload with width of 5 mm. This feature is very unique regarding the profiles of the other controllers (see Figs. 5.28–5.31 again). For instance, the $\mathcal{H}_\infty$ controllers without explicit consideration of model uncertainties and the PID one reveal the feature that RMSE decreases with payload increase.

On the other hand, the control performances of the SMC-RSSI2 enhanced with respect to robustness is extremely superior to those of the other controllers as seen in Fig. 6.19. It should be noted that the order of RMSEs the SMC-RSSI2 is less than those of the other controllers by a factor of 10. This result demonstrates that how easily one can increase the robustness of such a SMC-based controller by only increasing the

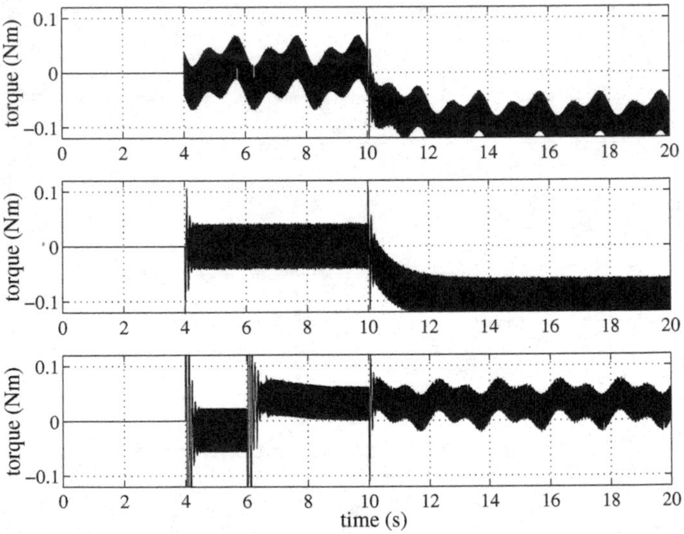

Fig. 6.20 Control torques of the joint 1 with the SMC-RSSI (0.0018 deg, $k_x = 12$, $k_z = 45$) for the respective problems in the case of double-frequency oscillation with the nominal payload of 5 mm; *top* attitude control in local coordinates, *middle* attitude control in global coordinates, *bottom* position control in global coordinates

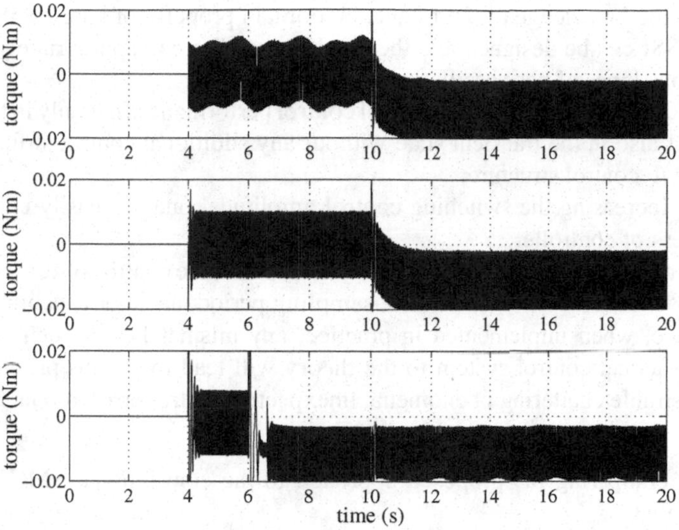

Fig. 6.21 Control torques of the joint 2 with the SMC-RSSI (0.0018 deg, $k_x = 12$, $k_z = 45$) for the respective problems in the case of double-frequency oscillation with the nominal payload of 5 mm; *top* attitude control in local coordinates, *middle* attitude control in global coordinates, *bottom* position control in global coordinates

switching control amplitude as k_x and k_z, which is the most advantageous property of the SMC schemes. However, the larger switching control amplitude means the more difficulty in its implementation in practice. Then, let us see the control torque inputs of the SMC-RSSI2 in the nominal payload case in Figs. 6.20 and 6.21. By comparing the results to those of the SMC-RSSI1 in Figs. 6.12 and 6.13, it is seen that the extremely large amplitude is required for the SMC-RSSI2 according to the amplitude of k_x and k_z. Therefore, practical implementation of the SMC-RSSI2 might lead to undesirable chattering phenomena and/or unexpected control error due to the mismatching capability of actuators to the theory.

6.6 Conclusions

In this chapter, we have explained our proposed SMC-based control approach, i.e., the SMC-RSSI, conducted stability analyses with respect to the RSSI, and demonstrated its capabilities by simulations. In order to compare this approach to the proposed $\mathcal{H}_\infty$-control-based one in the foregoing chapters, the method for demonstration is consistent with that for the $\mathcal{H}_\infty$-control-based one. Consequently, we have concluded that, with respect to comparison to the $\mathcal{H}_\infty$-control-based approach:

- Without the knowledge of the frequency-domain property of base oscillation, an SMC-RSSI can be designed and the single controller can appropriately perform for various types of base oscillations;
- The SMC-RSSI can provide successful control performances not only in the steady state, but also in the transient state without any additional control structure such as a TDOF control structure;
- By only increasing the switching control amplitude, one can easily enhance the robustness of controller;
- However, the SMC-RSSI is considerably sensitive to the control system conditions such as sensor resolution and noise, sampling period, actuator capability, and so on. Hence, when implemented in practice, any mismatches of such conditions of the practical control system to the theory will lead to serious problems such as undesirable chattering phenomena, unexpected control error, too large control inputs.

Moreover, comparing the proposed approach to the conventional SMC, we have confirmed that

- The SMC-RSSI can achieve the superior transient control performance such as rapid response without overshooting, and can suppress large control input.

In this monograph, we have not presented experimental demonstration for the SMC-based approach, because the present experimental OBM is not capable of appropriately implementing the SMC-RSSI in terms of the sensor resolution, sampling period, switching ability of the actuators. This point is one the next steps to be taken in future works.

Chapter 7
Base Oscillation Estimation via Multiple $\mathcal{H}_\infty$ Filters

Abstract This chapter presents a complementary technique to support the aforementioned control methodologies which have assumed that accurate measurements of the base oscillation can be available for the control system. Here we present an estimation algorithm of the base oscillation assuming a low-cost rate gyro sensor. The heart of this algorithm is a method of selectively combining multiple $\mathcal{H}_\infty$ filters according to an innovation-based criterion. We introduce this algorithm and demonstrate its estimation performance by simulations using the Bretschneider oscillation in Chap. 2. The results show that among the multiple filters appropriate ones can be selected with the innovation-based criterion, and thus the estimation algorithm is effective.

7.1 Introduction

In the foregoing chapters, both the proposed control methods have been addressed and demonstrated assuming that accurate measurements of the base oscillation are available. In practice, the accuracy of measurements depends on sensing systems used. In general, a high-precision sensing system such as a fiber optical gyro-based system requires expensive cost. Therefore, we have been involved in development of ship oscillation estimation algorithm using a relatively low-cost rate gyro sensor in order to support and enhance the assumption for the control methods.

In this chapter, an estimation algorithm of ship oscillation with a low-cost rate gyro sensor is presented. In particular, situations in the presence of frequency-domain property variation of oscillation are considered, which necessarily occur due to changing the ship velocity and direction, i.e., the way of encountering the ocean waves.

There have been typical two research directions to deal with motion estimation methods for moving objects. One of which is to combine measurements from the multiple sensors, so-called "sensor fusion technique" [55, 81], and the other one is to develop or improve model-based filtering algorithm such as the Kalman filter [23, 29, 54, 63, 66, 100, 107]. As examples of the former works, Leavitt et al. [55] and Sanca et al. [81] have proposed methods of compensating drift and integral errors

M. Toda, *Robust Motion Control of Oscillatory-Base Manipulators*,
Lecture Notes in Control and Information Sciences 463,
DOI 10.1007/978-3-319-21780-2_7

of rate gyro sensors using together other types of sensors, such as an accelerometer and inclinometer for the purpose of developing an attitude estimation method. On the other hand, as an example of the latter works, to predict wave-induced oscillatory ship motions, Küchler et al. [54] have presented a motion estimation method for a crane-vessel using a fast Fourier transform (FFT)-based state-equation model and the conventional Kalman filter. Further, Liu et al. [63] have developed a method of improving autoregressive model by the Kalman filter to estimate ship motions.

Motivated by [54] and adding a new viewpoint of variation of ship oscillation frequency, we have developed an estimation method of ship oscillation via a low-cost rate gyro sensor, which incorporates periodic updates of the FFT-based model and a method of selectively combining multiple $\mathcal{H}_\infty$ and Kalman filters, in order to enhance adaptiveness and robustness of estimation for a model variation. In general, such variation will deteriorate the estimation performance particularly when relying on a model-based estimation algorithm such as the Kalman filter. The method of combining multiple filters according to an innovation-based criterion is the key of our estimation method. The Kalman filtering algorithm, which is one of the most popular algorithms [23, 63, 100, 112], is to effectively reduce Gaussian white noises, while the H_∞ filtering one [33, 85], which has been becoming popular, e.g., [6, 27, 47, 49, 59, 78, 106], can accommodate norm-bounded deterministic noises, and therefore is expected to be robust against model errors. Our method of combining filters is to utilize the advantage of each filter as much as possible.

This chapter is organized as follows. In Sect. 7.2, we briefly explain the conventional Kalman and $\mathcal{H}_\infty$ filters and introduce our proposed method. Section 7.3 presents an ocean wave spectrum model and a ship motion model for demonstrations. Then, Sect. 7.4 demonstrates estimation performances of our method by simulations. The last section will give concluding remarks.

7.2 Estimation Algorithm

This section presents our proposed estimation algorithm. Figure 7.1 shows an outline of the algorithm, which consists of the processes of acquisition of a measurement of the angular velocity, periodic FFT-based model updates, selectively combining $\mathcal{H}_\infty$ and Kalman filters, integral with drift error compensation, and estimation of the base inclination angle, T_{FFT} is the FFT period, Δt_{FFT} is the interval to update the FFT-based model.

How to build the FFT-based state-equation model in Fig. 7.1 is addressed in Sect. 7.2.1. Then, Sects. 7.2.2 and 7.2.3 briefly review the conventional discrete Kalman filter and $\mathcal{H}_\infty$ filter, respectively. Next, Sect. 7.2.4 presents the method of selectively combining multiple filters, which is the heart of our methodology. Finally, to output the angle not velocity, the method of integral of velocity with drift error compensation is explained in Sect. 7.2.5.

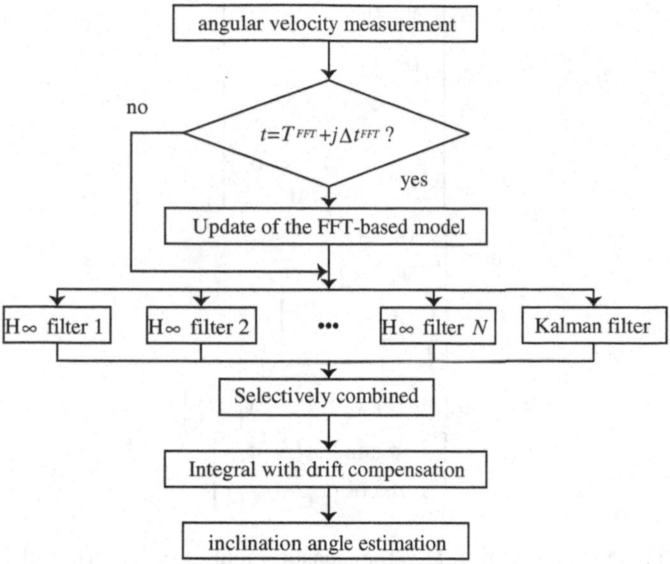

Fig. 7.1 Outline of the estimation algorithm; $0 \leq j \in \mathbb{Z}$: sampling index, T_{FFT}: FFT period, Δt_{FFT}: update period

7.2.1 FFT-Based Linear State-Equation Model

In accordance with the notations in the foregoing chapters, let $q_b(t)$ denote the base inclination angle to be estimated, and thus $\dot{q}_b(t)$ is the angular velocity to be measured via a gyro sensor. Further, let $\dot{q}_b(t)$ be decomposed into multiple sine waves called "modes" using FFT. From all the modes, the n dominant modes with the n largest amplitudes are chosen and then $\dot{q}_b(t)$ is approximated by

$$\dot{q}_b(t) \approx \sum_{i=1}^{n} a_i \sin(\omega_i t + \phi_i), \tag{7.1}$$

where a_i, ω_i and ϕ_i are the amplitude, angular frequency, and phase of the ith mode, respectively. Further, among the chosen modes, the angular frequency of the most dominant mode, i.e., with the largest amplitude, is denoted by ω_{max}, which is utilized in the processes of selectively combining multiple filters and integral with drift error compensation as in the sequel.

The dominant modes are incorporated into a state-space equation as in the following:

$$\dot{\mathbf{x}}(t) = A\boldsymbol{x}(t), \tag{7.2}$$

$$y(t) = C\boldsymbol{x}(t) + v(t), \tag{7.3}$$

$$
\begin{aligned}
A &= \begin{bmatrix} A_1 & 0 & \cdots\cdots & 0 \\ 0 & A_2 & \ddots & \vdots \\ \vdots & \ddots & \ddots & \ddots & \vdots \\ \vdots & & \ddots & \ddots & 0 \\ 0 & \cdots\cdots & 0 & A_n \end{bmatrix}, \\
A_i &= \begin{bmatrix} 0 & 1 \\ -\omega_i^2 & 0 \end{bmatrix}, \\
C &= \begin{bmatrix} C_1 & C_2 & \cdots & C_n \end{bmatrix}, \\
C_i &= \begin{bmatrix} 1 & 0 \end{bmatrix}, \\
\boldsymbol{x}(t) &= \begin{bmatrix} \boldsymbol{x}_1^T(t) & \boldsymbol{x}_2^T(t) & \cdots & \boldsymbol{x}_n^T(t) \end{bmatrix}^T, \\
\boldsymbol{x}_i(t) &= \begin{bmatrix} a_i \sin(\omega_i t + \phi_i) \\ \omega_i a_i \cos(\omega_i t + \phi_i) \end{bmatrix},
\end{aligned}
$$

where $\boldsymbol{x}(t)$ is the state vector, $y(t)$ is the measurement of $\dot{q}_b(t)$, $v(t)$ is the measurement error, which is assumed to be a Gaussian white noise, $i = 1, 2, \ldots, n$. This model is time invariant, and is updated every Δt_{FFT}.

Then, this continuous-time model expressed by (7.2) and (7.3) is discretized as follows:

$$\boldsymbol{x}_{k+1} = \bar{A}\boldsymbol{x}_k, \tag{7.4}$$

$$y_k = C\boldsymbol{x}_k + v_k, \tag{7.5}$$

$$z_k = C\boldsymbol{x}_k, \tag{7.6}$$

where $\bar{A} = \exp(A\Delta t)$, Δt is the sampling period, $[\cdot]_k := [\cdot](t_k)$, $t_k = k\Delta t$, $0 \leq k \in \mathbb{Z}$, $z_k = \dot{q}_{bk}$ to be estimated. Every Δt_{FFT}, the model is updated and k is reset to 0. Note that this formulation does not include any system noise not as in the conventional one.

7.2.2 Kalman Filtering Algorithm

The Kalman filtering algorithm for the system in (7.4)–(7.6) is given by

$$\hat{\boldsymbol{x}}_{k+1/k} = \bar{A}\hat{\boldsymbol{x}}_{k/k}, \tag{7.7}$$

$$\hat{\boldsymbol{x}}_{k/k} = \hat{\boldsymbol{x}}_{k/k-1} + K_k^f(y_k - \hat{z}_{k/k-1}), \tag{7.8}$$

$$\hat{z}_{k+1/k} = C\hat{\boldsymbol{x}}_{k+1/k}, \tag{7.9}$$

$$\hat{z}_{k/k} = C\hat{\boldsymbol{x}}_{k/k}, \tag{7.10}$$

where $\boldsymbol{x}_k$ is the state vector, K_k^f is the Kalman gain, $\hat{[\cdot]}_{i/j}$ is the estimate of $[\cdot]_i$ at the jth time point, $\hat{z}_{k/k}$ is the kth output of the filter. The Kalman gain K_k^f is obtained by

$$K_k^f = P_k C^T (C P_k C^T + R)^{-1}, \tag{7.11}$$

where R is the variance of v_k, which is also updated every Δt_{FFT}, P_k is the covariance matrix of the estimation error $[\boldsymbol{x}_k - \hat{\boldsymbol{x}}_{k/k-1}]$, which is recursively updated by the recurrence formula:

$$P_{k+1} = \bar{A}(P_k - K_k^f C P_k)\bar{A}^T. \tag{7.12}$$

The initial conditions $\hat{\boldsymbol{x}}_{0/-1}$, P_0 and R are set as follows:

$$\hat{\boldsymbol{x}}_{0/-1} = \boldsymbol{x}_0, \tag{7.13}$$

$$P_0 = \frac{1}{n_{FFT}} \sum_{k=-n_{FFT}+1}^{0} \tilde{\boldsymbol{x}}_k \tilde{\boldsymbol{x}}_k^T, \tag{7.14}$$

$$R = \frac{1}{n_{FFT}} \sum_{k=-n_{FFT}+1}^{0} v_k^2, \tag{7.15}$$

$$\tilde{\boldsymbol{x}}_k = \boldsymbol{x}_k - \frac{1}{n_{FFT}} \sum_{j=-n_{FFT}+1}^{0} \boldsymbol{x}_j, \tag{7.16}$$

$$v_k = y_k - \sum_{i=1}^{N_{FFT}} a_i \sin(\omega_i t_k + \phi_i), \tag{7.17}$$

where $n_{FFT} = T_{FFT}/\Delta t$ set to be $\in \mathbb{Z}^+$, $\boldsymbol{x}_k$, $k = -n_{FFT}+1, -n_{FFT}+2, \ldots, 0$ is obtained by the FFT-based model (7.4).

The advantage of the Kalman filter is to effectively reduce Gaussian white noise, under the condition that the state-space model by $\bar{A}$ is available. However, if the variation of the ship oscillation frequency is large and the current $\bar{A}$ is no longer available, the estimation performance using the Kalman filter will deteriorate.

7.2.3 $\mathcal{H}_\infty$ Filtering Algorithm

First, the suboptimal problem formulation of $\mathcal{H}_\infty$ filter is expressed by

$$\sup_{x_{0,v}} \frac{\sum_k (z_k - \hat{z}_k)^2}{(\boldsymbol{x}_0 - \hat{\boldsymbol{x}}_0)^T P_0^{-1} (\boldsymbol{x}_0 - \hat{\boldsymbol{x}}_0) + R^{-1} \sum_k v_k^2} < \gamma^2. \tag{7.18}$$

In our approach, γ is utilized as a filter design parameter, i.e., preparing multiple $\mathcal{H}_\infty$ filters with different γ's. Note that the initial covariance matrix of estimation error P_0 and the covariance matrix of measurement error R are used as the weighting function matrices in the above criterion in 7.18 instead.

Then, the $\mathcal{H}_\infty$ filtering algorithm for the problem based on the prescribed system is similar to that of the Kalman filter. The only difference is that the Kalman gain is replaced by the $\mathcal{H}_\infty$ filter gain H_k^f:

$$H_k^f = P_k C^T (C P_k C^T + R)^{-1}, \tag{7.19}$$

$$P_{k+1} = \bar{A} P_k \Psi_k^{-1} \bar{A}^T, \tag{7.20}$$

$$\Psi_k = I + (R^{-1} - \gamma^{-2}) C^T C P_k, \tag{7.21}$$

where I is the identity matrix, γ is the design parameter. In particular, if $\gamma \to \infty$, the $\mathcal{H}_\infty$ filtering algorithm coincides with the Kalman filtering one. Note that it is in general difficult to appropriately set γ. Hence, we employ multiple filters with different γ's, which is the key point in the proposed algorithm. The number of the employed $\mathcal{H}_\infty$ filters is denoted by N.

The existence condition for the $\mathcal{H}_\infty$ filter is that $P_k > 0$ and $\Psi_k P_k^{-1} > 0$ $(k = 0, 1, 2, \ldots)$. Further, one can easily confirm that $R^{-1} - \gamma^{-2} \geq 0$ and $P_0 > 0$ can be its sufficient condition. Therefore, we define $\beta_\gamma := (R^{-1} - \gamma^{-2})R$ $(0 \leq \beta_\gamma \leq 1)$, and use β_γ as the design parameter instead of γ in the proposed method. In this problem, R is a scaler. β_γ's are set as $(m/N)^2$ $(m = 0, 1, \ldots, N)$. In particular, when $m = N$, the filter is the Kalman filter.

The advantage of the $\mathcal{H}_\infty$ filter is to effectively reduce bounded-norm deterministic noises and therefore is expected to be robust against the model-error, i.e., the variation of ship oscillation frequency.

7.2.4 Method of Selectively Combining Multiple $\mathcal{H}_\infty$ and Kalman Filters

To overcome the problem of model deterioration due to the ship oscillation frequency change, we have proposed a method of selectively combining multiple $\mathcal{H}_\infty$ and Kalman filters, which aims at utilizing the advantage of each filter as much as possible. The selection of filters are based on a criterion related to innovations, i.e., gaps between measurements and estimates. For each filter, the Eqs. (7.8)–(7.10) can be rewritten as follows:

$$\hat{z}_{k/k} = \alpha_k y_k + (1 - \alpha_k)\hat{z}_{k/k-1}, \tag{7.22}$$

$$\alpha_k = C P_k C^T \left(C P_k C^T + R \right)^{-1}, \tag{7.23}$$

which displays the role of the defined criterion denoted by α. In this form, α is interpreted as a weight of contribution to the final estimate of the measurement y_k compared with that of $\hat{z}_{k/k-1}$. Thus, the smaller α implies that $\hat{z}_{k/k-1}$ based on the current model is the more reliable. Importantly, α will never reflect the actual innovations but has been scheduled in advance once the filter has been designed. Therefore, we have focused on this point and considered that if the predetermined α is "in a sense" in accordance with the actual innovations then the current model is appropriate and available. Note that CP_kC^T is the variance of estimation error of $\hat{z}_{k/k-1}$ and $CP_kC^T + R$ is the variance of the innovation $\nu_k = y_k - \hat{z}_{k/k-1}$. Hence, α_k can be approximated by

$$\alpha'_k := \left(\overline{\nu_i^2} - R\right)\left(\overline{\nu_i^2}\right)^{-1}, \tag{7.24}$$

$$\overline{\nu_k^2} := \frac{1}{n_m} \sum_{i=k-n_m+1}^{k} \nu_i^2, \tag{7.25}$$

where n_m is the rounded integer of $2\pi/(\omega_{max}\Delta t)$, ω_{max} is the angular frequency with the maximum amplitude in (7.1).

As mentioned above, α_k is predefined and does not contain any feedback of the measurements once the model has been updated, whereas α'_k reflects the recent measurements in the form of innovations, which includes the information on the model variation between the model updates. Therefore, $\alpha_k \geq \alpha'_k$ can be interpreted that the filter actually works as expected or better than expected. Thus, using this condition as an criterion, we select the filters at each time step and combine the respective estimates by averaging them, which is central idea of our proposed methodology.

7.2.5 Integral with Drift Error Compensation

To acquire inclination angles, estimates of angular velocities via the algorithm are numerically integrated by the trapezoidal rule. $\hat{q}'_b$ denotes the integral. The integral is in general influenced by drift errors of the gyro sensor. Hence, the following drift error compensation algorithm is used:

$$\hat{q}_{bj} = \begin{cases} \hat{q}'_{bj} & (j < T_{FFT}/\Delta t + n_m - 1) \\ \hat{q}'_{bj} - \overline{\hat{q}'_{b}}_j & (j \geq T_{FFT}/\Delta t + n_m - 1) \end{cases}, \tag{7.26}$$

$$\overline{\hat{q}'_{b}}_j = \frac{1}{n_m} \sum_{m=j-n_m+1}^{j} \hat{q}'_{bm}, \tag{7.27}$$

where n_m is utilized as in the process of filter selection as prescribed before.

7.3 Ship Oscillation Motions

This section addresses ship oscillation motion for demonstration by simulation, which is subject to variation of ship velocity. Specifically, an ocean wave spectrum model and ship dynamical models are presented.

7.3.1 Ocean Wave Spectrum Model

As an ocean wave spectrum model, the Bretschneider spectrum [22] is employed with a slight modification. See Sect. 2.4.2 again for the spectrum. In the demonstrations, we use the wave spectrum model not as motion but forces driving the ship dynamics. By linearly shifting ω_i's as $G_w\omega_i$'s (G_w is a constant multiplier) and linearly magnifying the amplitudes $A_{\omega i}$'s as $G_A A_{\omega i}$ (G_A is a constant gain), two types of wave models denoted by Model 1 and Model 2 are considered as shown in Table 7.1. Let f_w denote the normalized wave force expressed by

$$f_w(t) = \Sigma_{i=1}^{n_\omega} G_A A_{\omega i} \sin(G_w\omega_i t + \phi_i). \tag{7.28}$$

Additionally, variation of ocean wave frequency due to ship velocity variation is considered. Figure 7.2 depicts the variation rate of the frequency during the whole simulation period of 160 s, which is divided into five periods and labeled with Periods 1–5, respectively. Figure 7.3 displays the respective wave force profiles.

7.3.2 Ship Dynamical Model

As an illustrative example of an equation of ship motion [69, 108], that of ship-rolling motion is considered as in the following:

$$\ddot{q}_b(t) + 0.50\dot{q}_b(t) + 5.0q_b(t) + cq_b^3(t) = f_w(t) \tag{7.29}$$

where $q_b(t)$ is the inclination angle, c is the constant coefficient representing the nonlinearity, $f_w(t)$ is the normalized wave force mentioned above. When $c = 0$, the natural damped frequency of this model is 0.71π rad/s, which corresponds to

Table 7.1 Ocean wave models

	Model 1	Model 2
Peak frequency (rad/s)	0.72π	0.36π
G_w	3	1.5
G_A	3	2

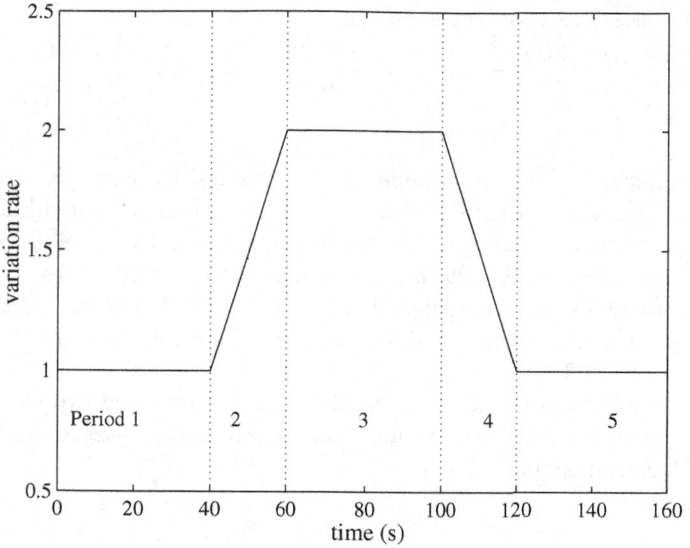

Fig. 7.2 Variation process of the ocean wave frequency

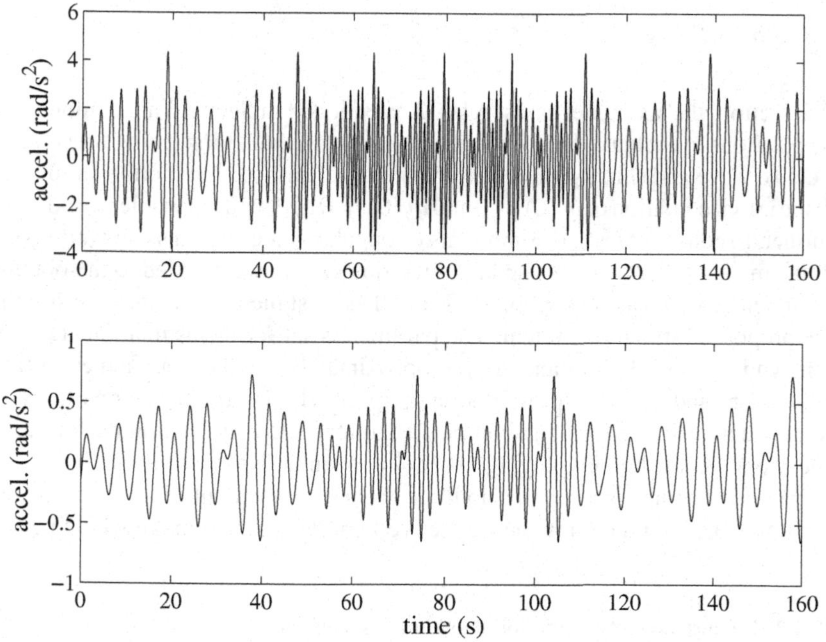

Fig. 7.3 Normalized external forces (acceleration); *top* Model 1, *bottom* Model 2

Table 7.2 Nonlinear parameter c of ship dynamical model

Nonlinear parameter	Model A	Model B	Model C
c	0	200	−30

the peak frequency as shown in Table 7.1. This natural frequency is equal to the peak frequency in Periods 1 and 5 with Model 1 and in Period 3 with Model 2. The constant coefficient c is related to the ship model nonlinearity, namely, with $c = 0$ (7.29) is linear, with $c > 0$ (7.29) has a nonlinear stiffness of the hard-spring type, and $c < 0$ provides a soft-spring type nonlinearity. According to the three cases of c, three dynamical models named Model A, B and C are considered respectively as presented in Table 7.2.

All the combinations of the three ship dynamical models and two wave models will be demonstrated in simulations, which are referred to as Cases A1, A2, B1, B2, C1, and C2, corresponding to the model indices.

7.4 Simulations

7.4.1 Conditions

In this section, the six model cases, A1–C2, mentioned in Sect. 7.3 are demonstrated by simulations to evaluate the proposed estimation method. First, being based on the respective wave and ship dynamical models, we performed ship motion simulations with the initial condition of $q_b(0) = 0$ and $\dot{q}_b(0) = 0$ and with the whole period of 160 s. Further, to construct test measurements of angular velocity, Gaussian white noises with the mean of 0°/s and the standard deviation of 8°/s were added to the obtained angular velocity data as sensor noises. Then, these test measurements were supplied to the proposed estimation system. The parameters set for the estimation algorithm are shown in Table 7.3. T_{FFT} and Δt_{FFT} approximately equal to four and eight times the natural period of the ship oscillation, respectively. Under those conditions, we have conducted simulations and compared the respective estimation performances of the proposed method employing 10 $\mathcal{H}_\infty$ and single Kalman filters, the integral with drift error compensation with no filter, from the viewpoint of root mean square error (RMSE) of estimation in the respective Periods 1–5, and in the whole period.

Table 7.3 Parameters for the estimation algorithm in simulations

Δt	T_{FFT}	Δt_{FFT}	n	N
0.1 s	11.2 s	22.4 s	5	10

Δt sampling period, T_{FFT} FFT period, Δt_{FFT} model update period, n number of modes, N number of $\mathcal{H}_\infty$ filters

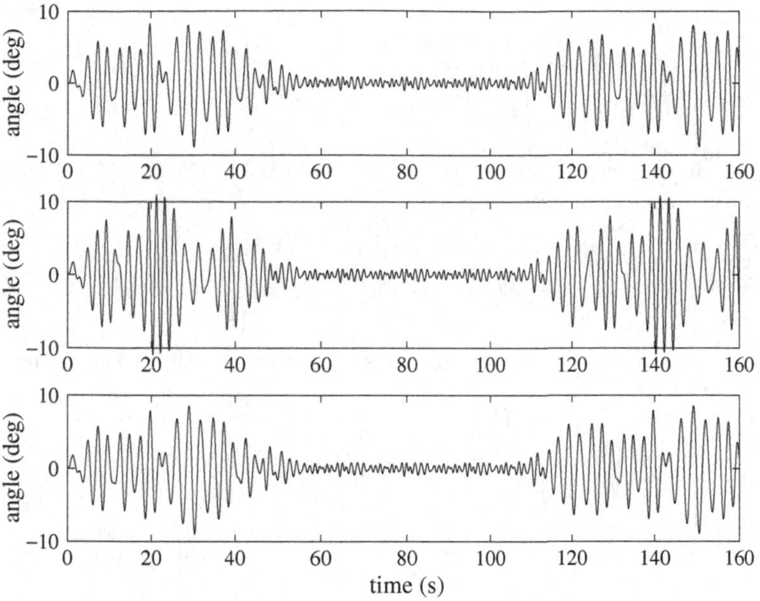

Fig. 7.4 Ship inclination angles; *top* Case A1, second from the *top* B1, *bottom* C1

7.4.2 Results

First let us see the simulation results of ship motion in the respective cases as shown in Figs. 7.4 and 7.5. Figure 7.4 compares the three ship dynamical models with the wave model Model 1. The features of the profiles are similar despite the differences in the dynamical models, where at $t = 50-110$ (Period 2–Period 4) the amplitude has dramatically decayed. Comparing the dynamical models, in Periods 1 and 5 the amplitude of Case B1 (hard-spring) is the largest, that of A1 (linear) is the second, and that of C1 (soft-spring) is the smallest. On the other hand, in Fig. 7.5 (for Model 2), quite different profiles from those in the case of Model 1 can be observed, where the amplitudes do not vary much according to the periods. These results are due to the property of low-pass filter of the ship dynamical models. Namely, waves with frequencies much greater than the natural damped frequency near 0.71π rad/s is difficult to be passed whereas waves with frequencies less than that can be passed easily. In terms of comparison of the models, the tendency is the opposite from that for Model 1. In Periods 1 and 5, the amplitude of Case C2 (soft-spring) is the largest, that of A2 (linear) is the second, and that of B2 (hard-spring) is the smallest.

Next, we examine the simulation results in the form of root mean square errors (RMSEs) of estimation which are shown in Figs. 7.6, 7.7, 7.8, 7.9, 7.10 and 7.11 for the respective model cases. In each figure, the circle represents the RMSE of the proposed method, * is for that of the best single filter, + is for that of the worst single filter, the triangle is for that of the integral with drift compensation without a

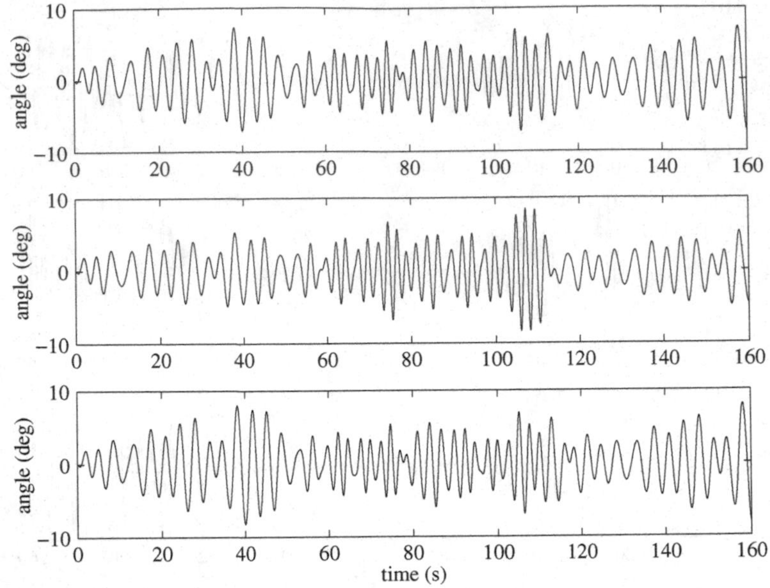

Fig. 7.5 Ship inclination angles; *top* Case A2, second from the *top* B2, *bottom* C2

Fig. 7.6 Simulation results of RMSE for Case A1 (linear ship model); Period 6 represents all the periods 1–5, ○: proposed filter, *: the best filter, +: the worst filter, △: integral compensated

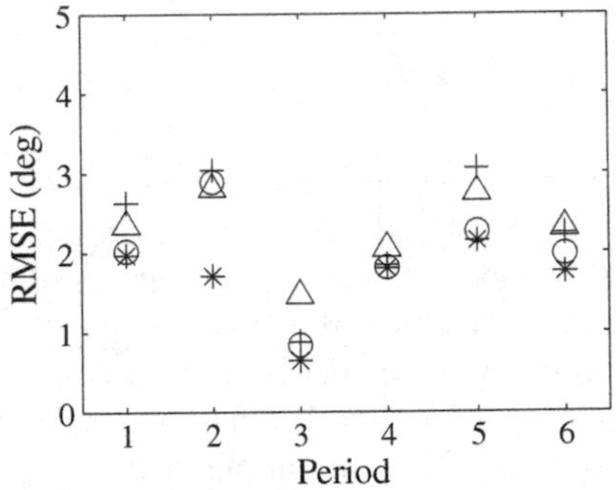

Fig. 7.7 Simulation results of RMSE for Case B1 (nonlinear ship model); Period 6 represents all the periods 1–5, ○: proposed filter, *: the best filter, +: the worst filter, △: integral compensated

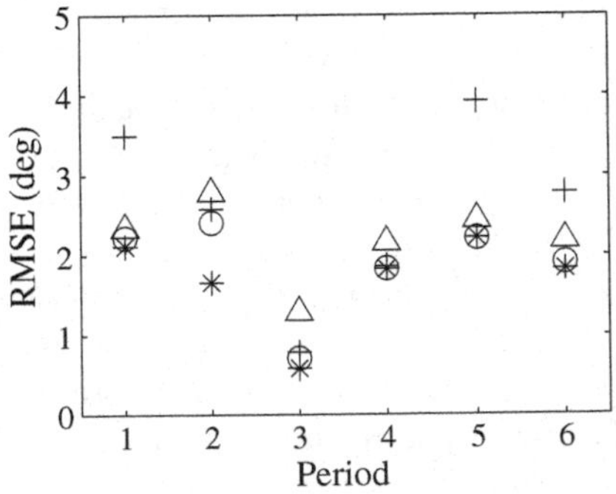

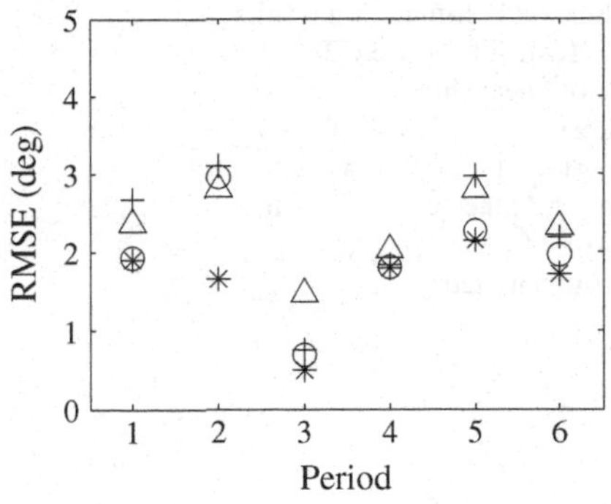

Fig. 7.8 Simulation results of RMSE for Case C1 (nonlinear ship model); Period 6 represents all the periods 1–5, ○: proposed filter, *: the best filter, +: the worst filter, △: integral compensated

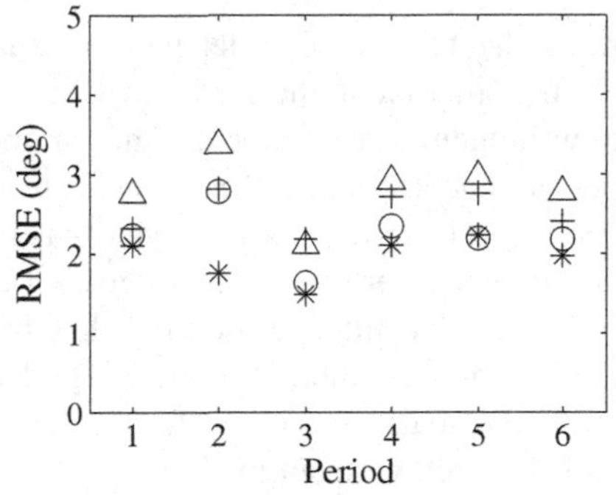

Fig. 7.9 Simulation results of RMSE for Case A2 (linear ship model); Period 6 represents all the periods 1–5, ○: proposed filter, *: the best filter, +: the worst filter, △: integral compensated

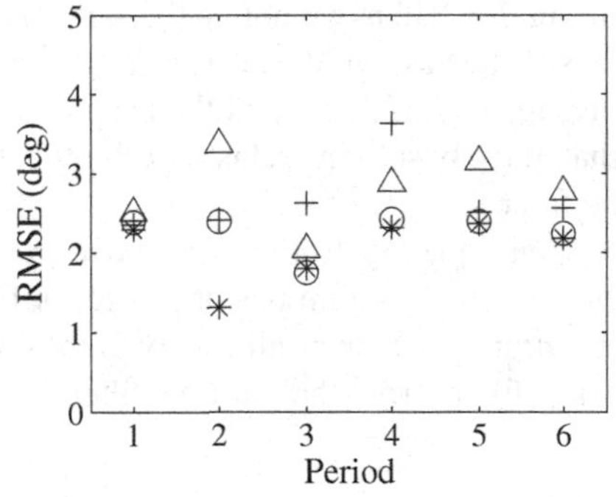

Fig. 7.10 Simulation results of RMSE for Case B2 (nonlinear ship model); Period 6 represents all the periods 1–5, ○: proposed filter, *: the best filter, +: the worst filter, △: integral compensated

filter. Period 6 means the whole period (Periods 1–5). Let us see Figs. 7.6, 7.7 and 7.8 which display the results in the case of Model 1 and compare the corresponding ship oscillation data in Fig. 7.4. First, by focusing the best single filter (*), compare the RMSEs and the corresponding ship oscillation carefully. Then, it is seen that the RMSEs are proportional to the amplitudes of ship oscillation, and are not influenced by the situations of time-varying and time-invariant oscillation frequency, nor by the nonlinearities in the ship dynamical models. Whereas the other RMSEs exhibit similar tendencies but which are not as clear as that of the best single filter. It should be

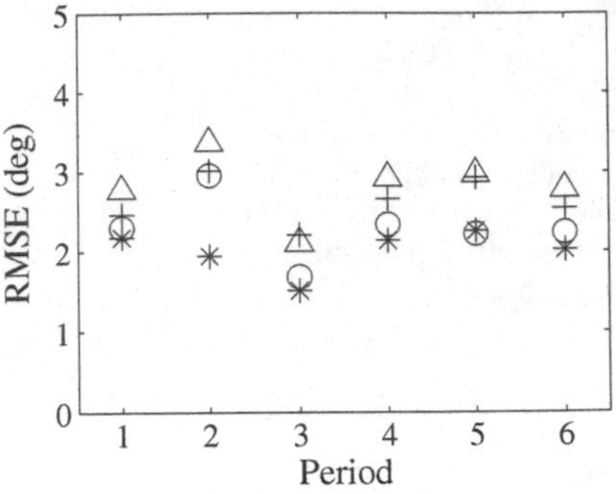

Fig. 7.11 Simulation results of RMSE for Case C2 (nonlinear ship model); Period 6 represents all the periods 1–5, ○: proposed filter, *: the best filter, +: the worst filter, △: integral compensated

noted that the best filter depends on situations and cannot be determined in advance. On the other hand, the performances of the worst filter is quite poor, which implies how influential the filter design parameter β_γ will be to the resulting estimation performance. Then, let us investigate the performances of the proposed method. From the figures, it is observed that in most cases the proposed method exhibit similar good performances to those of the best filter, except in Period 2, which has proven that the filter selection method base on the criterion α is effective. Comparing the proposed method with the integral with drift error compensation, the proposed method reveals superior performance.

Moreover, we examine Figs. 7.9, 7.10 and 7.11 for the case of Model 2. Similarly, we compare the RMSEs in those figures and the corresponding ship oscillation data in Fig. 7.5. Then, we notice the same argument can be applied to this case of Model 2 as in the case of Model 1. The performance of the best filter is only influenced by the signal amplitude, and the proposed method can gives very near performance to that of the best filter and is superior to the integral with drift error compensation with no filter.

Consequently, from all the above arguments, we have confirmed that the proposed estimation algorithm performs successfully in the presence of oscillation frequency variation and even nonlinearity in the ship dynamics by effectively selecting appropriate filters from the multiple filters.

7.5 Conclusions

In this chapter, an estimation method of ship oscillation in the presence of oscillation frequency variation has been presented, assuming that only an angular velocity is measured with a low-cost rate gyro sensor. This method incorporates periodic updates of the FFT-based model and the method of selectively combining multiple $\mathcal{H}_\infty$ and Kalman filters according to the innovation-based criterion. To evaluate the estimation performance of the proposed method, simulations have been conducted with consideration of the variation of ship oscillation frequency and nonlinearities in the ship dynamics. The results have proven that the proposed method is effective and robust under such difficult conditions, and justified the idea of combining filters aiming at utilizing the advantage of each filter. However, there still remains a room for improvement with respect to the filter selection method, which the results in Period 2 in the simulations have suggested. In fact, we are planning adding another type of criterion to the current innovation-based one.

References

1. Abdallah C, Dawson D, Dorato P, Jamshidi M (1991) Survey of robust control for rigid robots. IEEE Control Syst Mag 11:24–30
2. Alves RM, Battista RC (1999) Active control of heave motion for tlp type offshore platform. In: Proceedings of the international offshore and polar engineering conference, Brest, France, pp 332–338
3. Apkarian P, Chretien JP, Gahinet P, Biannic JM (1993) μ synthesis by $\mathcal{D}$-$\mathcal{K}$ iterations with constant scaling. In: Proceedings of the american control conference, pp 3192–3196
4. Balas GJ, Doyle JC, Glover K, Packard A, Smith R (1995) μ-Analysis and synthesis toolbox user's guide. MUSYN Inc. and The Math Works Inc, Minneapolis
5. Bandyopadhyay B, Deepak F, Kim K-S (2009) Sliding mode control using novel sliding surfaces. LNCIS, vol 392. Springer, Heidelberg
6. Batista P, Silvestre C, Oliveira P (2008) Kalman and $\mathcal{H}_\infty$ optimal filtering for a class of kinematic systems. In: Proceedings of the 17th world congress in IFAC, pp 12528–12533
7. Battista RC (1999) Active control of heave motion for tlp type offshore platform under random waves. Smart Structures and Materials 1999: Smart Systems for Bridges. Structures and Highways, Newport Beach, USA, pp 184–193
8. Becker N, Grimm WM (1988) On L_2- and L_∞-stability approaches for the robust control of robot manipulators. IEEE Trans Autom Control 33:118–122
9. Bowden KF (1983) Physical oceanography of costal waters. Ellis Horwood Limited, Chichester
10. Chida Y, Yamaguchi Y, Soga H, Kida T, Yamaguchi I, Sekiguchi T (1996) On-orbit attitude control experiments for ETS-VI–I-PD and two-degree-of-freedom $\mathcal{H}_\infty$ control–. In: Proceedings of the IEEE conference on decision and control, pp 486–488
11. Choi SB, Park DW (1994) Moving sliding surfaces for fast tracking control of second-order dynamic systems. ASME J Dyn Syst Meas Control 116:154–158
12. Choi SB, Park DW, Jayasuriya S (1994) A time-varying sliding surface for fast and tracking control of second-order dynamic systems. Automatica 30:899–904
13. Cloutier JR, D'Souza CN, Mracek CP (1996) Nonlinear regulation and nonlinear H_∞ control via the state-dependent Riccati equation technique: part 1, theory. In: Proceedings of the international conference on nonlinear problems in aviation and aerospace, pp 117–130
14. Desoer CA, Vidyasagar M (1975) Feedback systems: input-output properties. Academic, New York
15. Do KD, Pan J (2008) Nonlinear control of an active heave compensation system. Ocean Eng 35:558–571

M. Toda, *Robust Motion Control of Oscillatory-Base Manipulators*,
Lecture Notes in Control and Information Sciences 463,
DOI 10.1007/978-3-319-21780-2

16. Doyle JC, Glover K, Khargonekar PP, Francis BA (1989) State-space solutions to standard $\mathcal{H}_\in$ and $\mathcal{H}_\infty$ control problems. IEEE Trans Autom Control 34:831–847
17. Doyle J, Packard A, Zhou K (1991) Review of LFTs, LMIs, and μ. In: Proceedings of the IEEE conference on decision and control, pp 1227–1232
18. Dubowsky S, Papadopoulos E (1991) The kinematics, dynamics, and control of free-flying and free-floating space robotic systems. IEEE Trans Robot Autom 9(5):531–543
19. Fadali MS, Yaz E (1995) Stability robustness and robustification of the exact linearization method of robotic manipulator control. In: Proceedings of the IEEE conference on decision and control, pp 1624–1629
20. Fallaha CJ, Saad M, Kanaan HY, Al-Haddad K (2011) Sliding-mode robot control with exponential reaching law. IEEE Trans Ind Electron 58(2):600–610
21. Fang Y, Wang P, Sun N, Zhang Y (2014) Dynamics analysis and nonlinear control of an offshore boom crane. IEEE Trans Ind Electron 61(1):414–427
22. Fossen TI (1994) Guidance and control of ocean vehicles. Wiley, Chichester
23. Fossen TI, Perez T (2009) Kalman filtering for positioning and heading control of ships and offshore rigs. IEEE Control Syst Mag 32–46
24. Francsis B (1987) A course in $\mathcal{H}_\infty$ control theory. Springer, New York
25. Furuta K (1990) Sliding mode control of a discrete system. Syst Control Lett 14:145–152
26. Furuta K (1993) VSS type self-tuning control. IEEE Trans Ind Electron 40(1):37–44
27. Gérard B, Souley H, Zasadzinski M, Darouach M (2011) H_∞ filter for bilinear systems using LPV approach. Sig Process 91:1168–1181
28. Glover K, Doyle JC (1988) State-sspace formulae for all stabilizing controllers that satisfy and H_∞-norm bound and relations to risk sensitivity. Syst Control Lett 11:167–172
29. Godhavn JM (1998) Adapting tuning of heave filter in motion sensor. In: Proceedings of the OCEANS, pp 174–178
30. Gu D-W, Petkov PH (2013) Robust control desing with MATLAB®, 2nd edn. Springer, London
31. Gu K, Zohdy MA, Loh NK (1990) Necessary and sufficient conditions of quadratic stability of uncertain linear systems. IEEE Trans Autom Control 35:601–604
32. Ha QP, Rye DC, Durrant-Whyte HF (1999) Fuzzy moving sliding mode control with application to robotic manipulators. Automatica 35(4):607–616
33. Hassibi B, Sayed AH, Kailath T (1996) Linear estimation in Krein spaces–part II: applications. IEEE Trans Autom Control 41(1):34–49
34. El-Hawary F, Mbamalu GAN (1996) Dynamic heave compensation using robust estimation techniques. Comput Electr Eng 22(4):257–273
35. Hootsmans NAM, Dubowsky S (1991) Large motion control of mobile manipulators including vehicle suspension characteristics. In: Proceedings of the IEEE international conference on robotics and automation, Sacramento, USA, pp 2336–2341
36. Hootsmans NAM, Dubowsky S, Mo PZ (1992) The experimental performance of a mobile manipulator control algorithm. In: Proceedings of the IEEE international conference on robotics and automation, pp 1948–1954
37. Hosoe S, Zhang T (1988) An elementary state space approach to RH^∞ optimal control. Syst Control Lett 11:369–380
38. Huang S, Vassalos D (1995) Optimal control of the heave motion of marine cable subsea-unit systems. In: Proceedings of the international offshore and polar engineering conference, The Hague, Netherlands, pp 203–208
39. Isram S, Liu XP (2011) Robust sliding mode control for robot manipulators. IEEE Trans Ind Electron 58(6):2444–2453
40. Isidori A (1995) Nonlinear control systems. Springer, Berlin
41. Itkis U (1976) Control systems of variable structure. Wiley, New York
42. Iwamura T, Toda M (2013) Motion control of an oscillatory-base manipulator using sliding mode control via rotating sliding surface with variable-gain integral control. In: Proceedings of the american control conference, Washington DC, USA, pp 5762–5767

43. Iwasaki T (1994) Robust performance analysis for systems with norm-bounded time-varying structured uncertainty. In: Proceedings of the american control conference, Maryland, pp 2343–2347
44. Iwasaki T, Skelton RE (1994) All controllers for general $\mathcal{H}_\infty$ control problem: LMI existence conditions and state space formulas. Automatica 30(8):1307–1317
45. Johansen TA, Fossen TI, Sagatun SI, Nielsen FG (2003) Wave synchronizing crane control during water entry in offshore moonpool operations-experimental results. IEEE J Ocean Eng 28(4):720–728
46. Joshi J, Desrochers AA (1986) Modeling and control of a mobile robot subject to disturbances. In: Proceedings of the IEEE international conference on robotics and automation, pp 1508–1513
47. Karimi HR (2009) Robust H_∞ filter design for uncertain linear systems over network with network-induced delays and output quantization. Model Identif Control 30(1):27–37
48. Khalil HK (2002) Nonlinear systems, 3rd edn. Prentice Hall, Upper Saddle River
49. Kharrati H, Khanmohammadi S (2008) Genetic algorithm combined with H∞ filtering for optimizing fuzzy rules and membership function. J Appl Sci 8(19):3439–3445
50. Kimura H (1981) Robust stabilizability for a class of transfer functions. IEEE Trans Autom Control 27:783–793
51. Kimura H (1989) Conjugation, interpolation and model-matching in H_∞. Int J Control 49:269–307
52. Korde UA (1998) Active heave compensation on drill-ships in irregular waves. Ocean Eng 25(7):541–561
53. Kreutz K (1989) On manipulator control by exact linearization. IEEE Trans Autom Control 34:763–767
54. Küchler S, Mahl T, Neupert J, Schneider K, Sawodny O (2011) Active control for an offshore crane using prediction of the vessel's motion. IEEE-ASME Trans Mechatron 17(2):297–309
55. Leavitt J, Sideris A, Bobrow JE (2011) Accurate tilt sensing with linear model. IEEE Sens J 11(10):2301–2309
56. Levant A (1993) Sliding order and sliding accuracy in sliding model control. Int J Control 58(6):1247–1263
57. Lew JY, Moon SM (1999) Acceleration feedback control of compliant base manipulators. In: Proceedings of the american control conference, pp 1955–1958
58. Lew JY, Moon SM (2001) A simple active damping control for compliant base manipulators. IEEE/ASME Trans Mechatron 6:305–310
59. Li W, Jia Y (2010) H-infinity filtering for a class nonlinear discrete-time systems based on unscented transform. Signal Process 90:3301–3307
60. Liang Y-W, Xu S-D, Liaw D-C, Chen C-C (2008) A study of T-S model-based SMC scheme with application to robot control. IEEE Trans Ind Electron 55(11):3964–3971
61. Lin J, Huang ZZ (2007) A hierarchical fuzzy approach to supervisory control of robot manipulators with oscillatory bases. Mechatoronics 17:589–600
62. Lin J, Lin CC, Lo H-S (2009) Psedo-inverse jacobian control with grey relational analysis for robot manipulators mounted on oscillatory bases. J Sound Vib 326:421–437
63. Liu Z, Yang Q, Guo Z, Li J (2011) An improved autoregressive method with Kalman filtering theory for vessel motion prediction. Intell Eng Syst 4(4):11–18
64. Jimenez-Lozano N, Goodwine B (2010) Nonlinear disturbance decoupling for a nonholonomic mobile robotic manipulation platform. In: Proceedings of the IEEE international conference on control. automation, robotics and vision, pp 1530–1535
65. Magee DP, Book WJ (1995) Filtering micro-manipulator wrist commands to prevent flexible base motion. In: Proceedings of the american control conference, pp 924–928
66. Mahony R, Hamel T, Pflimlin J-M (2008) Nonlinear complementary filters on the special orthogonal group. IEEE Trans Autom Control 53(5):1203–1218
67. Marconi L, Isidori A, Serrani A (2002) Autonomous vertical landing on an oscillating platform: an internal-model based approach. Automatica 38:21–32

68. Megretski A (1993) Necessary and sufficient conditions of stability: a multi-loop generalization of the circle criterion. IEEE Trans Autom Control 38:753–756
69. Morrall A (1980) The Gaul disaster: an investigation into the loss of a large stern trawler. Trans R Inst Nav Archit 123:391–416
70. Nenchev DN, Yoshida K, Vichitkulsawat P, Konno A, Uchiyama M (1997) Experiments on reaction null-space based decoupled control of a flexible structure mounted manipulator system. In: Proceedings of the IEEE international conference on robotics and automation, pp 2528–2534
71. Neupert J, Mahl T, Haessig B, Sawodny O, Schneider K (2008) A heave compensation approach for offshore cranes. In: Proceedings of the american control conference, Seattle, USA, pp 538–543
72. Packard A, Doyle J (1993) The complex structured singular value. Automatica 29:71–109
73. Ngo QH, Hong K-S (2012) Sliding-mode antisway control of an offshore container crane. IEEE-ASME Trans Mechatron 17(2):201–209
74. Papadopoulos E, Dubouwsky S (1991) On the nature of control algorithms for free-floating space manipulators. IEEE Trans Robot Autom 7(6):750–758
75. Pernebo L (1981) An algebraic theory for the design of controllers for linear multivariable systems–part I: structure matrices and feedforward design. IEEE Trans Autom Control 26:171–182
76. Poolla K, Tikku A (1995) Robust performance against time-varying structured perturbations. IEEE Trans Autom Control 40(9):1589–1602
77. Qu Z, Dawson DM (1996) Robust tracking control of robot manipulators. IEEE Press, New Jersey
78. Rho H, Kang Y, Hyun S, Kim H (2010) Discrete H_∞ estimator design of unknown input: game-theoretic approach. IEEE Trans Autom Control 55(7):1674–1688
79. Rotea MA, Iwasaki T (1994) An alternative to the $D - K$ iteration?. In: Proceedings of the american control conference, pp 53–57
80. Sampei M, Mita T, Nakamichi M (1990) An algebraic approach to H_∞ output feedback control problems. Syst Control Lett 14:13–24
81. Sanca A, Ferreira JP, Javier P (2012) A real-time attitude estimation scheme for hexarotor micro aerial vehicle. ABCM Symp Ser Mechatron 5:1160–1166
82. Sato M, Toda M (2009) Motion control of an oscillatory-base manipulator in the global coordinates. In: proceedings of the IEEE international conference control and automation, New Zealand, pp 349–354
83. Sato M, Toda M (2015) Robust motion control of an oscillatory-base manipulator in a global coordinate system. IEEE Trans Ind Electron 62(2):1163–1174
84. Sato M, Toda M (2014) Estimation of ship oscillation subject to ship speed variation using an algorithm combining H_∞ and Kalman Filters. In: Proceedings IEEE international conference control and automation, Taichung, Taiwan, pp 929–934 (IEEE Trans Ind Electron 62(2):1163–1174 (2015))
85. Shaked U, Theodor Y (1992) H_∞-optimal estimation: a tutorial. In: Proceedings of the IEEE international conference decision and control. Tucson, Arizona, pp 2278–2286
86. Shamma JS (1994) Robust stability with time-varying structured uncertainty. IEEE Trans Autom Control 39:714–724
87. Sharon A, Hardt D (1984) Enhancement of robot accuracy using end-point feedback and a macro-micro manipulator system. In: Proceedings of the american control conference, pp 1836–1842
88. Skaare B, Egeland O (2006) Parallel force/position crane control in marine operations. IEEE J Ocean Eng 31(3):599–613
89. Spong MW (1992) On the robust control of robot manipulators. IEEE Trans Autom Control 37:1782–1786
90. Spong MW, Vidyasagar M (1987) Robust linear compensator design for nonlinear robotic control. IEEE J Robot Autom 3:345–351
91. Spong MW, Vidyasagar M (1989) Robot dynamics and control. Wiley, New York

92. Stein G, Doyle JC (1991) Beyond singular values and loop shapes. J Guid 14:5–16
93. Tikku A, Poolla K (1993) Robust performance against slowly-varying structured perturbations. In: Proceedings of the IEEE conference on decision and control, pp 990–995
94. Toda M (1999) Robust control for mechanical systems with oscillating bases. In: Proceedings of the IEEE international conference on systems man cybernetics, vol 2. Tokyo, Japan, pp 878–883
95. Toda M (2004) An $\mathcal{H}_\infty$ control-based approach to robust control of mechanical systems with oscillatory bases. IEEE Trans Robot Autom 20(2):283–296
96. Toda M (2004) A unified approach to control of mechanical systems with a flexible structure. In: Proceedings of the international symposium on robotics and automation, Mexico, pp 313–319
97. Toda M (2007) A unified approach to robust control of flexible mechanical systems. In: Proceedings of the IEEE international conference on decision and control, New Orleans, USA, pp 5787–5792
98. Toda M (2010) A unified approach to robust control of flexible mechanical systems using $\mathcal{H}_\infty$ control powered by PD control. In: Shafiei SE (ed) Advanced Strategies for Robot Manipulators, Sciyo, Rijeka, Croatia, ch 13. pp 273–286
99. Torres MA, Dubowsky S, Pisoni AC (1994) Path-planning for elastically-mounted space manipulators: experimental evaluation of the coupling map. In: Proceedings of the IEEE international conference on robotics and automation, pp 2227–2233
100. Triantafyllou MS, Bodson M, Athans M (1983) Real time estimation of ship motions using Kalman filtering techniques. IEEE J Eng OE-8(1):9–20
101. Utkin VI (1977) Variable structure system with sliding mode. IEEE Trans Autom Control 22(2):212–221
102. Utkin VI (1992) Sliding mode in control optimization. Springer, New York
103. Utkin VI (1993) Sliding mode control design principles and applications to electric drives. IEEE Trans Ind Electron 40(1):23–36
104. Vidyasagar M, Kimura K (1986) Robust controllers for uncertain linear multivariable systems. Automatica 22:85–94
105. Wakasa Y, Yamamoto Y (1999) Control system design considering a tradeoff between evaluated uncertainty ranges and control performance. Asian J Control 1:49–57
106. Wang QC, Li J, Zhang MX, Yang CH (2011) H-infinity filter based particle filter for maneuvering target tracking. Prog Electromagn Res B 30:103–116
107. Wang XF, Zhao XR, Yang XJ (2006) Research of motion resolving and filtering algorithm of a ship's three-freedom motion simulation platform based on LabVIEW. Instrum Sci Technol 48:149–153
108. Wright JHG, Marshfield WB (1979) Ship roll response and capsize behavior in beam seas. Trans R Inst Nav Archit 122:129–144
109. Yamada Y, Hara S (1998) Global optimization for $\mathcal{H}_\infty$ control with constant diagonal scaling. IEEE Trans Autom Control 43:191–203
110. Yamamoto Y, Yun X (1994) Coordinating locomotion and manipulation of a mobile manipulator. IEEE Trans Autom Control 39(6):1326–1332
111. Yang J, Li S, Yu X (2013) Sliding-mode control for systems with mismatched uncertainties via a disturbance observer. IEEE Trans Ind Electron 60(1):160–169
112. Yun X, Aparicio C, Bachmann ER, McGhee RB (2005) Implementation and experimental results of a quaternion-based kalman filter for human body motion tracking. In Proceedings of the IEEE international conference on robotics and automation, Barcelona, pp 317–322
113. Zames G (1981) Feedback and optimal sensitivity: Model reference transformations, multiplicative seminorms, and approximate inverses. IEEE Trans Autom Control 26:301–320
114. Zames G, Francis BA (1983) Feedback, minimax sensitivity, and optimal robustness. IEEE Trans Autom Control 28:585–601
115. Zhou K, Doyle JC, Glover K (1995) Robust and optimal control. Prentice Hall, New Jersey

Index

M. Toda, *Robust Motion Control of Oscillatory-Base Manipulators*,
Lecture Notes in Control and Information Sciences 463,
DOI 10.1007/978-3-319-21780-2

Printed in the United States
By Bookmasters